**KUMON MATH WORKBOOKS**

# Word Problems
## Workbook I

## Table of Contents

KUM○N

# 1 Review 1

Date / /

Name

■ The Answer Key is on page 88.

**1** James is making a paper chain from different colored paper strips.    10 points
He joins the links in the following pattern: red, blue, red and blue, and so on.
Counting from the first link, what color will the 23rd link be?

⟨Ans.⟩ _____

**2** Mary drank $\frac{4}{5}$ liters ( ℓ ) of juice and left $2\frac{3}{5}$ liters ( ℓ ) of juice in the carton.    10 points
How many liters of juice were in the carton before Mary drank the juice?

⟨Ans.⟩ _____ ℓ

**3** John's family drinks $\frac{7}{9}$ ℓ of milk a day.    10 points
How many liters do they drink in 4 days?

⟨Ans.⟩ _____ ℓ

**4** A saline solution has a weight of 1.12 kg per liter.    10 points
What is the weight of the saline solution of 0.75 ℓ?

⟨Ans.⟩ _____ kg

**5** A car can drive 35.5 km using 3.5 ℓ of gasoline.    10 points
How many kilometers can the car drive using 1 ℓ of gasoline?

(Let's round off the quotient and calculate to the tenths place.)

⟨Ans.⟩ about _____ km

**6** Robert's weight is 73.6 pounds, Tiffany's weight is 66.5 pounds, and Johnny's weight is 85.8 pounds. Find the average of their weight.

10 points

〈Ans.〉 _____ pounds

**7** Patricia's class has 30 students. Among them, there are 18 girls. What percentage of the class are girls?

10 points

〈Ans.〉 _____ %

**8** Michael purchased 280 apples. At the end of the day, he sold 35% of his apples. How many apples did he sell?

10 points

〈Ans.〉 _____ apples

**9** William bought canned goods at $2.40 per can. He wants to increase the price per can so that he makes a 25% profit. What will the selling price per can be?

10 points

〈Ans.〉 $ _____

**10** Jennifer bought 1 apple and 3 oranges for $5.10 and Linda bought 5 oranges and 1 apple for $7.50. What is the price of 1 apple and 1 orange, respectively?

10 points

〈Ans.〉 Apple  $ _____

〈Ans.〉 Orange  $ _____

Date
/ /

Name

■ The Answer Key is on page 88.

**1** The bus bound for Northtown departs every 15 minutes and the bus for Easttown departs every 18 minutes from the Central Station. Both buses start at 9:30 AM each morning. When is the next time both buses will depart from the Central Station at the same time?

10 points
...........

⟨Ans.⟩ _____ AM

**2** David had $3\frac{1}{2}$ gallons of paint. From that, he used $1\frac{1}{3}$ gallons to paint the roof of the kennel. How many gallons of paint are left?

10 points
...........

⟨Ans.⟩ _____ gallons

**3** Elizabeth's family drinks $2\frac{1}{2}$ ℓ of apple juice in 6 days. How many liters do they drink per day?

10 points
...........

⟨Ans.⟩ _____ ℓ

**4** A kettle holds 2.24 times as much water as a pot. The amount of water the pot can hold is 2.5 ℓ. How many liters of water can the kettle hold?

10 points
...........

⟨Ans.⟩ _____ ℓ

**5** A rectangle with a length of 3.2 m has an area of 14.4 m². What is the width of the rectangle in meters?

10 points
...........

⟨Ans.⟩ _____ m

**6** If an average orange weighs 75 g, how many oranges would weigh 4.5 kg?

〈Ans.〉 _____ oranges

10 points

**7** The West Elementary School has 540 students, and its gymnasium area is 450 m². On the other hand, the East Elementary School has 728 students, and its gymnasium area is 560 m². When all the students entered the gymnasium in each elementary school, which elementary school gymnasium is more crowded?
Let's compare by number of people per 1 m².

10 points

〈Ans.〉 _____

**8** Richard's soccer team has 32 players. 12 of the players are fifth graders. What percentage of the team is made up of fifth graders?

10 points

〈Ans.〉 _____ %

**9** Barbara bought a notebook for $1.40 using 35% of the money she had. How much money did she start with?

10 points

〈Ans.〉 $ _____

**10** Susan disolves 20 g of salt into 230 g of water to create a saline solution. What percentage is the weight of the salt relative to the weight of the whole saline solution?

10 points

〈Ans.〉 _____ %

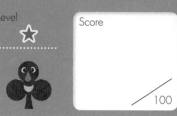

■ The Answer Key is on page 88.

★ **TRY!** — Fill in the blanks provided and solve for the answer.

**1** One bag contains $\frac{3}{4}$ kg of salt.

What is the weight of 5 bags of salt?

10 points

| The weight of salt in 1 bag | | The number of bags | | Overall weight | | Mixed fraction |
|---|---|---|---|---|---|---|
| | × | | = | | = | |

⟨Ans.⟩ _____ kg

**2** The weight of 3 cans of pineapple is $\frac{7}{10}$ kg.

What is the weight of 1 can of pineapple?

10 points

| Overall weight | | The number of cans | | The weight of 1 can |
|---|---|---|---|---|
| | ÷ | | = | |

⟨Ans.⟩ _____ kg

**3** Water runs into a natural spring at a rate of 7 ℓ per minute.

How many liters of water will run into the spring in $\frac{5}{6}$ of a minute?

10 points

| Amount that runs into the spring | | Time | | Overall amount | | Mixed fraction |
|---|---|---|---|---|---|---|
| | × | | = | | = | |

⟨Ans.⟩ _____ ℓ

**4** A 1 m piece of wire weighs $\frac{1}{8}$ kg.

What is the weight of $\frac{3}{7}$ m of this wire?

10 points

| Weight of 1 m | | Length of wire | | Overall weight |
|---|---|---|---|---|
| | × | | = | |

⟨Ans.⟩ _____ kg

**5** Joseph has 2 pounds of clay and Sarah has $\frac{1}{4}$ times the amount of

Joseph's clay. How much clay does Sarah have?

10 points

| Amount of Joseph's clay | | Proportion | | Amount of Sarah's clay |
|---|---|---|---|---|
| | × | | = | |

⟨Ans.⟩ _____ pounds

**6** Thomas is filling a bathtub with water. 10 points

10 ℓ of water will enter the bathtub in $\frac{3}{5}$ of a minute.

How many liters of water will enter the bathtub in a minute?

Amount of water     Time spent     Amount of water in a minute     mixed fraction

[   ] ÷ [   ] = [   ] = [   ]

⟨Ans.⟩ _____ ℓ

**7** Margaret has a piece of ribbon that is 7 feet long. 10 points

How many pieces of ribbon will she have if she cuts the original

ribbon every $\frac{1}{6}$ feet?

Length of ribbon     l length     Number of ribbons

[   ] ÷ [   ] = [   ]

⟨Ans.⟩ _____ pieces

**8** Charles's mother is a good cook. 10 points

Yesterday, she used $\frac{3}{5}$ ℓ of milk while cooking. Today she used 3 ℓ of milk.

How many times more milk did she use today than she used yesterday?

Amount of milk used today     Amount of milk used yesterday     Proportion

[   ] ÷ [   ] = [   ]

⟨Ans.⟩ _____ times

**9** Karen divided $\frac{3}{4}$ kg of salt into bags that each held $\frac{1}{12}$ kg. 10 points

How many bags of salt did she end up with?

The total weight of the salt     The weight of salt in l bag     The number of bags

[   ] ÷ [   ] = [   ]

⟨Ans.⟩ _____ bags

**10** A baby rabbit weighs $\frac{4}{5}$ of a kg and a baby cat weighs $\frac{3}{5}$ of a kg. 10 points

How many times more is the weight of the baby cat than the baby rabbit?

The weight of the baby cat     The weight of the baby rabbit     Proportion

[   ] ÷ [   ] = [   ]

⟨Ans.⟩ _____ times

# Fractions 2

Level ☆
Score          /100

Date    /    /

Name

■ The Answer Key is on page 88.

**1** Shauna and her friends divide $\frac{3}{4}$ kg of clay equally between 5 people. **10 points**
What is the weight of 1 person's clay?

〈Ans.〉_____ kg

**2** Christopher bought $1\frac{3}{4}$ kg of beans for $5 per 1 kg. **10 points**
How much did he pay for the beans?

〈Ans.〉 $ _____

**3** A large rectangular piece of cardboard is $1\frac{7}{9}$ m in length and $\frac{3}{4}$ m **10 points**
in width. What is the area of the piece of cardboard?

〈Ans.〉_____ m²

**4** Daniel harvested $3\frac{1}{5}$ kg of grapes from his farm. **10 points**
Nancy harvested $\frac{5}{6}$ times more than Daniel.
What is the weight of Nancy's grapes?

〈Ans.〉_____ kg

**5** An iron pipe weighs $\frac{3}{8}$ kg and is $\frac{2}{5}$ m long. **10 points**
How much would the iron pipe weigh if it was 1 m long?

〈Ans.〉_____ kg

**6** Lisa's classmates each brought $1\frac{2}{3}$ kg of rice and it came to 15 kg.
How many people brought rice?

10 points

⟨Ans.⟩ _____ people

**7** A piece of red tape is 10 feet long and a piece of white tape is $\frac{3}{2}$ feet long.
How many times longer is the red tape than the white tape?

10 points

⟨Ans.⟩ _____ times

**8** If Matthew divides $1\frac{1}{4}$ ounces of sugar into $\frac{5}{8}$ ounce bags,
how many bags will he have?

10 points

⟨Ans.⟩ _____ bags

**9** There are $1\frac{4}{5}$ ℓ of juice and $\frac{2}{7}$ ℓ of milk in a refrigerator.
How many times more is the amount of juice compared to the amount of milk?

10 points

⟨Ans.⟩ _____ times

**10** Marina has a piece of wire that is 6 m long and weighs $\frac{1}{5}$ kg per meter.
She uses 1 m of the wire in her work.
How much does the remaining wire weigh?

10 points

⟨Ans.⟩ _____ kg

# Fractions 3

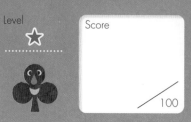

Level ☆    Score    /100

■ The Answer Key is on page 88.

**1** A toy robot can move $2\frac{2}{9}$ feet in 6 seconds.    10 points
How many feet per second does it move?

⟨Ans.⟩ _____ feet

**2** Betty's father bought a $3\frac{5}{6}$ m hose for \$3 per 1 m.    10 points
What was the total cost of the hose?

⟨Ans.⟩ \$ _____

**3** A toy car can move $1\frac{5}{7}$ m per minute.    10 points
How many meters would the car move in $1\frac{1}{6}$ minutes?

⟨Ans.⟩ _____ m

**4** A piece of wire is $1\frac{3}{7}$ m long and a piece of rope is $2\frac{5}{8}$ times the length    10 points
of the wire. How many meters is the rope?

⟨Ans.⟩ _____ m

**5** Anthony spread $3\frac{3}{4}$ kg of fertilizer on a field measuring $3\frac{1}{8}$ m².    10 points
How many square meters can he spread fertilizer per 1 kg?

⟨Ans.⟩ _____ m²

**6** A piece of tape is 18 m long. Donald cut it into $1\frac{1}{5}$ m long pieces.    10 points

How many pieces of tape did he make?

⟨Ans.⟩ _____ pieces

**7** A big bottle of water holds 3 ℓ and a small bottle of water    10 points

holds $2\frac{1}{4}$ ℓ.

How many times more water does the big bottle hold than the small bottle?

⟨Ans.⟩ _____ times

**8** Dorothy has a rope that is $7\frac{1}{2}$ m long.    10 points

How many pieces of rope will she have if she cuts the original rope every $1\frac{1}{4}$ m?

⟨Ans.⟩ _____ pieces

**9** Sandra bought $\frac{5}{6}$ m of tape at \$1.80 per meter, and she bought $\frac{5}{6}$ m    10 points

of tape at \$2.40 per meter.

How much did she pay for the tape?

⟨Ans.⟩ \$ _____

**10** Mark painted a section of a rectangular board with white paint as shown    10 points

by the figure below.

How many square meters is the painted area of the board?

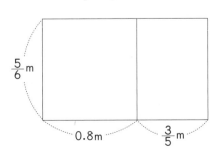

⟨Ans.⟩ _____ m²

# 6 Speed 1

Date / /

Name

■ The Answer Key is on page 89.

★ **TRY!** — Fill in the blanks provided and solve for the answer.

**1** **Paul drove his car 80 km in 2 hours.**
**What is his speed in kilometers per hour?**

10 points

( The speed in kilometers per hour = km/h )

| The distance | | The time | | The speed |
|---|---|---|---|---|
| | ÷ | | = | |

⟨Ans.⟩ _____ km/h

**2** **Ashley rode her bicycle 2,700 m in 15 minutes.**
**What is her speed in meters per minute?**

10 points

( The speed in meters per minute = m/min )

| The distance | | The time | | The speed |
|---|---|---|---|---|
| | ÷ | | = | |

⟨Ans.⟩ _____ m/min

**3** **Steven's horse ran 120 m in 8 seconds.**
**What is the speed in meters per second of his horse?**

10 points

( The speed in meters per second = m/sec )

| The distance | | The Time | | The speed |
|---|---|---|---|---|
| | ÷ | | = | |

⟨Ans.⟩ _____ m/sec

**4** **A sound travels 340 m per second through the air.**
**What is the speed in meters per minute of the sound?**

10 points

| The speed in meters per second | | | The speed in meters per minute |
|---|---|---|---|
| | × 60 | = | |

⟨Ans.⟩ _____ m/min

**5** **Kimberly's car can drive at 40 km per hour.**
**How many kilometers can she drive her car in 3 hours?**

10 points

| The speed | | The time | | The distance |
|---|---|---|---|---|
| | × | | = | |

⟨Ans.⟩ _____ km

**6** Donna and her friends walk a distance of 12 km during a hike.     10 points
How long will it take them to walk the 12 km at a pace of 3 km per hour?

The distance      The speed      The time

$$\boxed{\phantom{xxxx}} \div \boxed{\phantom{xxxx}} = \boxed{\phantom{xxxx}}$$

⟨Ans.⟩ _____ hours

**7** Andrew's father drove his car a distance of 54 miles in $\frac{3}{4}$ of an hour.     10 points

What is his speed in miles per hour?

The distance      The time      The speed

$$\boxed{\phantom{xxxx}} \div \boxed{\phantom{xxxx}} = \boxed{\phantom{xxxx}}$$

⟨Ans.⟩ _____ miles/h

**8** A train travels a distance of 60 km in 1 hour.     10 points
How many kilometers would the train travel in 45 minutes?

( Let's calculate the time as a fraction. )

The speed      The time      The distance

$$\boxed{\phantom{xxxx}} \times \boxed{\phantom{xxxx}} = \boxed{\phantom{xxxx}}$$

⟨Ans.⟩ _____ km

**9** A racecar can drive 5 km per minute.     10 points
What is the speed in kilometers per hour of the racecar?

The speed in kilometers per minute      The speed in kilometers per hour

$$\boxed{\phantom{xxxx}} \times \; 60 \; = \boxed{\phantom{xxxx}}$$

⟨Ans.⟩ _____ km/h

**10** A swallow can fly 1,500 m per minute.     10 points
How fast does the same swallow fly in the speed per second?

The speed per minute      The speed per second

$$\boxed{\phantom{xxxx}} \div \; 60 \; = \boxed{\phantom{xxxx}}$$

⟨Ans.⟩ _____ m/sec

  13

Date / /

Name

■ The Answer Key is on page 89.

**1** Kenneth ran around a pond with a perimeter of 72 m.
He ran 10 laps in 3 minutes.
What is Kenneth's speed in meters per second?

10 points

⟨Ans.⟩          m/sec

**2** A jet can fly at a speed of 900 miles per hour.
What is the speed of the jet in miles per minute?

10 points

⟨Ans.⟩          miles/min

**3** A train can travel a distance of 64 km per hour.
How many kilometers will it travel in 2 hours and 30 minutes?

10 points

⟨Ans.⟩          km

**4** Emily takes a bus that runs at a speed of 500 m per minute.
If the bus runs for 25 minutes, how many kilometers will it travel?

10 points

⟨Ans.⟩          km

**5** The distance from George's house to the station is 910 m.
Walking at a speed of 65 m per minute,
how long will it take George to walk to the station?

10 points

⟨Ans.⟩          minutes

**6** The distance between Carol's house and her uncle's house is 24 km. 10 points
What is her speed in kilometers per hour in order to get to her uncle's house in 40 minutes by motorcycle?

⟨Ans.⟩ _____ km/h

**7** Joshua ran 150 m around a pond in 24 seconds. 10 points
What is his speed per minute?

⟨Ans.⟩ _____ m/min

**8** Michelle got on the train and traveled 54 km from West Station to East Station at a speed of 720 m per minute. 10 points
How long did it take her to travel from West Station to East Station?

⟨Ans.⟩ _____ hour _____ minutes

**9** If an express train travels at a speed of 80 miles per hour for $1\frac{3}{4}$ hours, how far will it travel? 10 points

⟨Ans.⟩ _____ miles

**10** Kevin walks at a speed of $\frac{15}{4}$ km per hour. 10 points
It takes him 25 minutes to walk from his house to the station.
How many kilometers is the distance from his house to the station?

⟨Ans.⟩ _____ km

© Kumon Publishing Co., Ltd.    15

# Speed 3

Date            /            /

Name

■ The Answer Key is on page 89.

**1** Amanda's older brother rode his motorcycle to their uncle's house. **10 points**
He drove at a speed of 30 miles per hour and got to the house in 20 minutes.
How many miles is Amanda's house from her uncle's house?

⟨Ans.⟩ _____ miles

**2** Brian is driving his car on the highway at 72 miles per hour. **10 points**
He entered the highway at 9:40 AM and exited at 10:20 AM.
How many miles did Brian drive his car on the highway?

⟨Ans.⟩ _____ miles

**3** Melissa's horse can run at a speed of 12 m per second. **10 points**
How many minutes would it take her horse to run 21.6 km?

⟨Ans.⟩ _____ minutes

**4** Edward rode his bicycle 3 laps around a pond with a perimeter **10 points**
of 1.8 km at a speed of 6 m per second.
How many minutes did it take him to ride 3 laps around the pond?

⟨Ans.⟩ _____ minutes

**5** Deborah ran 100 m in 18 seconds. **10 points**
What is her speed per minute?

⟨Ans.⟩ _____ m/min

**6** Andre flew on an airplane that traveled 1,000 km in 1 hour and 20 minutes. What is the speed in kilometers per hour of this airplane?

⟨Ans.⟩ _____ km/h

**7** Stephanie drove her car to visit her uncle in a seaside town. It took her 2 hours and 15 minutes to drive 100 miles. What is the speed in miles per hour of her driving?

⟨Ans.⟩ _____ miles/h

**8** Timothy flew on an airplane that traveled at a speed of 480 km per hour. The airplane traveled 880 km. How long was Timothy's flight in hours and minutes?

⟨Ans.⟩ _____ hour _____ minutes

**9** Rebecca walked 2 km at a speed of $\frac{15}{4}$ km per hour. How many minutes did she walk for?

⟨Ans.⟩ _____ minutes

**10** Jason rode his bicycle at a speed of $\frac{55}{4}$ km per hour. How many minutes would it take him to bike a distance of 11 km?

⟨Ans.⟩ _____ minutes

Date / /    Name    Score /100

■ The Answer Key is on page 89.

**1** There are $x$ number of oranges in a bag and 4 oranges separately.    10 points
There are 12 oranges in total.
**How many oranges are in the bag?**

( Let's represent the number of oranges contained in the bag as $x$ and solve for an answer. )

$$x + 4 = 12$$
$$x = 12 - 4$$
$$x =$$

⟨Ans.⟩ _____ oranges

**2** There is 1 carton of milk.    10 points
Laura drank 3 ounces of milk and left 7 ounces of milk in the carton.
**How much milk was in the carton to start with?**

( Let's represent the amount of milk contained in the carton as $x$ and solve for an answer. )

$$x - 3 = 7$$
$$x = 7 + 3$$
$$x =$$

⟨Ans.⟩ _____ ounces

**3** Kyle bought 3 pencils for $1.80 at the store.    10 points
**How much did he pay for 1 pencil?**

( Let's represent the price of 1 pencil as $x$ and solve for an answer. )

$$x \times 3 = 1.80$$
$$x = 1.80 \div 3$$
$$x =$$

⟨Ans.⟩ $ _____

**4** There is a rhombus with a perimeter of 26 cm.    10 points
**What is the length of one of its sides?**

( Let's represent the length of one side as $x$ and solve for an answer. )

$$x \times 4 = 26$$
$$x = 26 \div 4$$
$$x =$$

⟨Ans.⟩ _____ cm

**5** Helen has a can full of oil.
She divides the oil into 5 bottles so each has the same amount of oil.
If each bottle now has 3.6 ounces of oil, how much oil was in the can?

10 points

( Let's represent the amount of oil as $x$ and solve for an answer. )

⟨Ans.⟩ _____ ounces

**6** A rectangle has a length of 9 cm and an area of 45 cm².
How many centimeters is the width of the rectangle?

10 points

( Let's represent the width of this rectangle as $x$ and solve for an answer. )

⟨Ans.⟩ _____ cm

**7** A parallelogram has a height of 8 cm and an area of 72 cm².
How many centimeters is the base of the parallelogram?

10 points

( Let's represent the base of this parallelogram as $x$ and solve for an answer. )

⟨Ans.⟩ _____ cm

**8** A rectangle has  a width of 6 cm and an area of 45 cm².
How many centimeters is the length of the rectangle?

15 points

( Let's represent the length of this rectangle as $x$ and solve for an answer as a decimal. )

⟨Ans.⟩ _____ cm

**9** A parallelogram has a base of 6 cm and an area of 33 cm².
How many centimeters is the height of this parallelogram?

15 points

( Let's represent the height of this parallelogram as $x$ and solve for an answer. )

⟨Ans.⟩ _____ cm

# Algebraic Expressions 2

Level  ★★

Date / /

Name

Score
/100

■ The Answer Key is on page 90.

**1** A rectangular prism has a width of 8 cm, a height of 4 cm, and a volume of 160 cm³. How many centimeters is the length of the rectangular prism?

10 points

( Let's represent the length of this rectangular prism as $x$ and solve for an answer. )

$$x \times 8 \times 4 = 160$$
$$x \times 32 = 160$$
$$x = 160 \div 32$$
$$x =$$

⟨Ans.⟩ _____ cm

**2** There are 4 boxes that each hold 5 pencils.
The total weight of the 4 boxes is 800 g.
How many grams does 1 pencil weigh?

10 points

( Let's represent the weight of 1 pencil as $x$ and solve for an answer. )

$$x \times 5 \times 4 = 800$$
$$x \times 20 = 800$$
$$x = 800 \div 20$$
$$x =$$

⟨Ans.⟩ _____ g

**3** There are 5 rows that each have 6 boxes of apple juice in a line.
The total price is $15.
What is the price of 1 apple juice?

10 points

( Let's represent the price of 1 apple juice as $x$ and solve for an answer. )

$$x \times 6 \times 5 = 15$$

⟨Ans.⟩ $ _____

**4** Nicolas bought 3 boxes of peaches with 4 peaches in each box.
The total price is $12.
What is the price of 1 peach?

10 points

( Let's represent the price of 1 peach as $x$ and solve for an answer. )

⟨Ans.⟩ $ _____

© Kumon Publishing Co., Ltd.

**5** Let's express the relationship between $x$ and $y$ in the following sentences.

5 points per question

( 1 ) Helen puts $x$ g of salt into a 30 g bag. Then, the total weight will be $y$ g.

⟨formula⟩  $30 + x = y$

( 2 ) Ryan bought an apple for $x$ and paid \$2. The change was \$$y$.

⟨formula⟩  $= y$

( 3 ) Sharon used 2.5 m from $x$ m of tape. The remaining tape is $y$ m long.

⟨formula⟩

( 4 ) Jacob bought $x$ books at \$2.60 per book. The total cost was \$$y$.

⟨formula⟩

( 5 ) A rectangle with a length of $x$ cm and a width of 9 cm has an area of $y$ cm².

⟨formula⟩

( 6 ) Cynthia put the same amount of $x$ ℓ oil in 5 bottles.
The amount of oil per bottle is $y$ ℓ.

⟨formula⟩

**6** Answer the questions about the relationship between the length of one side of an equilateral triangle and its other sides.

10 points per question

( 1 ) When the length of one side of an equilateral triangle is $x$ cm and the perimeter is $y$ cm, write an expression to find the perimeter.

⟨formula⟩

( 2 ) When $x$ is 4, find the value of $y$ and solve for the perimeter.

⟨Ans.⟩                    cm

( 3 ) When $y$ is 42, find the value of $x$ and solve for the length of one side.

⟨Ans.⟩                    cm

# Linear Equations 1
## (Solving Equations)

Date / /

Name

Score /100

■ The Answer Key is on page 90.

**1** **Solve each equation.**

3 points per question

( Show an improper fraction by a mixed fraction. )

(1) $x + 3 = 9$

$x = 9 - 3$
$x =$

⟨Ans.⟩ _____

(2) $x + 7 = 3$

⟨Ans.⟩ _____

(3) $6 + x = 0$

⟨Ans.⟩ _____

(4) $-5 + x = -5$

⟨Ans.⟩ _____

(5) $x - 0.8 = 4$

⟨Ans.⟩ _____

(6) $2 + x = 1.5$

⟨Ans.⟩ _____

(7) $x - 3.9 = 1.6$

⟨Ans.⟩ _____

(8) $-\dfrac{2}{3} + x = -5$

$x = -5 + \dfrac{2}{3}$
$x =$

⟨Ans.⟩ _____

(9) $x - 3 = -\dfrac{4}{5}$

⟨Ans.⟩ _____

(10) $6 + x = -\dfrac{3}{4}$

⟨Ans.⟩ _____

(11) $x - \dfrac{1}{3} = \dfrac{1}{2}$

⟨Ans.⟩ _____

(12) $\dfrac{1}{6} + x = \dfrac{1}{4}$

⟨Ans.⟩ _____

(13) $x - \dfrac{5}{6} = \dfrac{4}{3}$

⟨Ans.⟩ _____

(14) $-\dfrac{7}{4} + x = -\dfrac{3}{2}$

⟨Ans.⟩ _____

**2** **Solve each equation.**
( Show an improper fraction by a mixed fraction. )

( 1 )  $3x = 9$

$x = 9 \times \dfrac{1}{3}$

$x =$

⟨**Ans.**⟩ _____

( 2 )  $\dfrac{1}{3}x = -4$

$x = -4 \times 3$

$x =$

⟨**Ans.**⟩ _____

( 3 )  $-x = 7$

⟨**Ans.**⟩ _____

( 4 )  $\dfrac{1}{4}x = \dfrac{1}{3}$

⟨**Ans.**⟩ _____

( 5 )  $-3x = 24$

⟨**Ans.**⟩ _____

( 6 )  $-\dfrac{1}{6}x = 8$

⟨**Ans.**⟩ _____

( 7 )  $-3x = 8$

⟨**Ans.**⟩ _____

( 8 )  $\dfrac{1}{5}x = \dfrac{2}{3}$

⟨**Ans.**⟩ _____

( 9 )  $-\dfrac{2}{3}x = 6$

$x = 6 \times \left(-\dfrac{3}{2}\right)$

$x =$

⟨**Ans.**⟩ _____

(10)  $\dfrac{3}{4}x = \dfrac{2}{5}$

$x = \dfrac{2}{5} \times \dfrac{4}{3}$

$x =$

⟨**Ans.**⟩ _____

(11)  $-\dfrac{5}{6}x = 4$

⟨**Ans.**⟩ _____

(12)  $-\dfrac{1}{12}x = -\dfrac{2}{9}$

⟨**Ans.**⟩ _____

(13)  $\dfrac{4}{5}x = \dfrac{2}{3}$

⟨**Ans.**⟩ _____

(14)  $-\dfrac{7}{11}x = -\dfrac{5}{6}$

⟨**Ans.**⟩ _____

(15)  $-\dfrac{6}{7}x = 9$

⟨**Ans.**⟩ _____

(16)  $-\dfrac{7}{3}x = \dfrac{6}{5}$

⟨**Ans.**⟩ _____

# Linear Equations 2
## (Solving Equations)

**12**

Date    /    /

Name

Level ☆

Score          /100

■ The Answer Key is on page 90.

**1**  **Solve each equation.**
( Show an improper fraction by a mixed fraction. )

3 points per question

(1)  $3x - 4 = 5$

$3x = 5 + 4$

$3x = 9$

$x = 9 \times \dfrac{1}{3}$

$x =$

⟨Ans.⟩ _____

(2)  $4x - 2 = 10$

⟨Ans.⟩ _____

(3)  $-6x - 8 = 3$

⟨Ans.⟩ _____

(4)  $5x - 7 = 9$

⟨Ans.⟩ _____

(5)  $\dfrac{2}{3}x + 5 = -1$

⟨Ans.⟩ _____

(6)  $\dfrac{3}{5}x - 4 = 2$

⟨Ans.⟩ _____

(7)  $7x = 5x + (-10)$

$7x - 5x = -10$

$2x = -10$

$x = -10 \times \dfrac{1}{2}$

$x =$

⟨Ans.⟩ _____

(8)  $4x = x - 3$

⟨Ans.⟩ _____

(9)  $7 - 4x = 10x$

⟨Ans.⟩ _____

(10)  $-3x - 1 = 3x$

⟨Ans.⟩ _____

(11)  $\dfrac{1}{3}x - 5 = \dfrac{1}{2}x$

⟨Ans.⟩ _____

(12)  $-\dfrac{1}{4}x - 3 = -\dfrac{3}{2}x$

⟨Ans.⟩ _____

**② Solve each equation.**

(1) $7x - 8 = 4x + 10$

$7x - 4x = 10 + 8$

$3x = 18$

$x =$

〈Ans.〉 _____

(2) $3x - 6 = -7x + 10$

〈Ans.〉 _____

(3) $-2x - 8 = 2x + 12$

〈Ans.〉 _____

(4) $2x - 9 = -11x + 9$

〈Ans.〉 _____

(5) $7x - 6 = 4x + 1$

〈Ans.〉 _____

(6) $\frac{1}{2}x + 5 = \frac{1}{3}x + 2$

〈Ans.〉 _____

(7) $-\frac{2}{5}x + \frac{1}{4} = -\frac{1}{2}x + \frac{1}{3}$

〈Ans.〉 _____

(8) $\frac{5}{3}x - 3 = \frac{1}{2}x - \frac{3}{4}$

〈Ans.〉 _____

(9) $8x - 1 = 3x - 11$

$8x - 3x = -11 + 1$

$5x = -10$

$x =$

〈Ans.〉 _____

(10) $7x - 8 = 4x + \frac{1}{3}$

〈Ans.〉 _____

(11) $8x + 4 = \frac{1}{2}x + 12$

〈Ans.〉 _____

(12) $\frac{2}{5}x - 11 = -3x + \frac{3}{5}$

〈Ans.〉 _____

(13) $6x + \frac{1}{3} = \frac{3}{4}x + 2$

〈Ans.〉 _____

(14) $-\frac{1}{2}x - \frac{3}{7} = x + \frac{2}{9}$

〈Ans.〉 _____

(15) $2x - \frac{3}{5} = \frac{1}{2}x + \frac{1}{6}$

〈Ans.〉 _____

(16) $x + \frac{3}{4} = -\frac{1}{4}x + \left(-\frac{1}{8}\right)$

〈Ans.〉 _____

**13**

Level ☆☆

Date    /    /

Name

Score
/100

■ The Answer Key is on page 90.

---

## Pattern 1 — Number of pieces

● Gary bought several donuts at $ 0.80 per donut.
When the donuts were put into a box that cost $ 0.50,
the total price was $ 4.50.
How many donuts did he buy?

$0.80     $0.50

---

★ **TRY!** — Fill in the blanks provided and solve for the answer.    10 points per question

（1） What is the price if Gary bought $x$ donuts?

The price per donut         The number of donuts

[          ] × [          ] = [          ] dollars

（2） What is the price of the box?

[          ] dollars

（3） What is the total price?

[          ] dollars

（4） Find the equation for the price.

The price of the donuts         The price of the box         Total price

[          ] dollars + [          ] dollars = [          ] dollars

Solve the above equation.

⟨**Ans.**⟩ _____ donuts

---

**Hint**

First of all, it is important to think about what is being defined as $x$.

Let's represent the number being solved for by $x$.

---

**Key Point**

At first, decide what to put as $x$.
Secondly, find the relationship of equal quantities,
and make the equation using the equal sign $(=)$.

**1** **When Kathleen bought 1 notebook at \$1.40 and several pencils at \$0.60 per pencil. The total price was \$6.20. How many pencils did she buy? Fill in the blanks provided and solve for the answer.** 20 points

When Kathleen bought $x$ pencils,

The price of pencils ⇒ 0.60 × [　　　] dollars

The price of notebooks ⇒ [　　　] dollars

The total price ⇒ [　　　] dollars

The equation for the price,

The price of pencils + The price of notebooks [　　　　　] dollars = Total price [　　　] dollars

Solve the above equation. 〈Ans.〉 ______ pencils

**2** **When Amy bought several donuts at \$1.50 per piece and they were put into a box that cost \$0.50, the total price was \$11. How many donuts did she buy?** 20 points

〈Ans.〉 ______ donuts

**3** **Samantha bought several cakes at \$1.70 per piece and they were put into a box that cost \$0.25, the total price was \$10.45. How many cakes did she buy?** 20 points

〈Ans.〉 ______ cakes

Date / /

Name

■ The Answer Key is on page 90.

---

**Pattern 2 — Price**

● The total price of 8 oranges and 1 apple at $1.20 is double the total price of 1 orange and 1 pear at $1.80.
How much is the price of 1 orange?

doubled price of

$x   $1.20   =   $x   $1.80

---

★ **TRY!** — **Fill in the blanks provided and solve for the answer.**

15 points per question

（1） If the price of 1 orange is $x$,
what is the total price of 8 oranges and 1 apple?

The price of the orange

[＿＿＿] dollars ＋ [＿＿＿] dollars

The price of the apple

（2） What is the total price of 1 orange and 1 pear?

The price of the orange

[＿＿＿] dollars ＋ [＿＿＿] dollars

The price of the pear

（3） Write the equation for finding the price.

Total price of the orange and the apple

[＿＿＿] dollars ＝ 2 ( [＿＿＿] ) dollars

Total price of the orange and the pear

Solve the above equation.

⟨Ans.⟩ $ ＿＿＿＿＿

---

**Hint**

Be sure to check that the solution to the equation matches what the problem is asking.

For example, the prices are not fractions, or negative numbers.

So, the answer should not be in these forms.

Always check the solution from this point of view.

---

© Kumon Publishing Co., Ltd.

**Key Point**

When A is a multiple of B, it is expressed by the following equation.

"A is $x$ times B" $\Rightarrow$ "A $= x \times$ B"

---

**1** The total price of 6 pieces of drawing paper and 1 ballpoint pen at $0.70 is equal to the total price of a piece of drawing paper and a box of paints at $5.20.
How much is the price of a piece of drawing paper?
Fill in the blanks provided and solve for the answer.

15 points

If the price of a piece of drawing paper is $x$ dollars,
the total price of 6 pieces of the drawing paper and 1 ballpoint pen is,

$$6\ \boxed{\phantom{xxx}}\ \text{dollars}\ +\ \boxed{\phantom{xxx}}\ \text{dollars}$$

The total price of a piece of drawing paper and a box of paints is,

$$\boxed{\phantom{xxx}}\ \text{dollars}\ +\ \boxed{\phantom{xxx}}\ \text{dollars}$$

The equation for the total price is,

The total price of the drawing paper and the ballpoint pen

The total price of the drawing paper and the paints

$$\boxed{\phantom{xxx}}\ \text{dollars}\ =\ \boxed{\phantom{xxx}}\ \text{dollars}$$

Solve the above equation.

⟨Ans.⟩ $ _____

---

**2** The total price of 3 pencils and 1 protractor at $3 was 4 times the total price of 1 pencil and 1 eraser at $0.60.
What is the price of 1 pencil?

20 points

⟨Ans.⟩ $ _____

---

**3** The total price of 5 pencils and 1 protractor at $3 was twice the total price of 3 pencils and 2 erasers at $0.40 per one.
What is the price of 1 pencil?

20 points

⟨Ans.⟩ $ _____

# Linear Equations 5
## (Pattern 3)

Date    /    /

Name

Score
/100

■ The Answer Key is on page 91.

---

**Pattern 3 — Change**

● Eric had $5.
  When he bought 5 boxes of caramels and 1 box of gum
  at $0.40, the change was $0.60.
  What is the price of 1 box of caramels?

---

★ **TRY!** — **Fill in the blanks provided and solve for the answer.**

20 points per question

( 1 )  If the price of 1 box of caramels is $x$,
       what is the total price of 5 boxes of caramels and 1 box of gum?

The price of the caramels

5 [            ] dollars  +  [            ] dollars

The price of the gum

( 2 )  Write the equation to find the amount of change.

Amount of money paid

[            ] dollars  −  [            ] dollars  =  [            ] dollars

Total price                                    Change

Solve the above equation.

⟨Ans.⟩  $ _____

---

**Hint**

In this question,

considering "Total price" + "Change" = "Amount of money paid",
    $(5x + 0.4) + 0.6 = 5$

considering "Total price" = "Amount of money paid" − "Change",
    $5x + 0.4 = 5 - 0.6$

Both equations are correct.

There are a variety of ways equations can be written to solve the same problem.

---

**1** Shirley had $50.
When she bought 3 CDs and 1 calendar at $6, the change was $8.
What is the price of 1 CD?
Fill in the blanks provided and solve for the answer.

20 points

If the price of 1 CD is $x$,
what is the total price of 3 CDs and 1 calendar?

The price of the CDs

3 [       ] dollars + [       ] dollars

The price of the calendar

The equation to find the change is,

Amount of money paid — Total price

[                    ] dollars = [          ] dollars

Change

Solve the above equation.

⟨Ans.⟩ $ _____

**2** Calvin had $10.
When he bought 7 colored pencils and 2 notebooks at $1.30 per notebook,
the change was $1.10.
What is the price of 1 colored pencil?

20 points

⟨Ans.⟩ $ _____

**3** Keith had $10.
When he bought 5 colored pencils and 3 notebooks at $1.50 per notebook,
the change was $2.
What is the price of 1 colored pencil?

20 points

⟨Ans.⟩ $ _____

# Linear Equations 6
## (Pattern 4)

Score
/100

Date      /      /

Name

■ The Answer Key is on page 91.

---

## Pattern 4 — Two Quantities to Find

● Angela bought 10 stamps that cost $1 or $1.20.
Altogether, her total was $11.
How many of each $1 stamps and how many of
each $1.20 stamps did she buy?

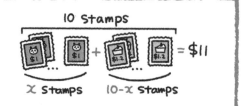

10 Stamps

$+$ $= \$11$

$x$ Stamps      $10-x$ Stamps

---

## ★ TRY! — Fill in the blanks provided and solve for the answer.

10 points per question

(1) If there are $x$ stamps that cost $1,
how many stamps cost $1.20?

10 − (stamps of $1)

[     ] stamps

(2) The price for $1 stamps      ⇒

The price for $1 stamps

[     ] dollars

The price for $1.20 stamps      ⇒

The price × The stamps

$1.20 ($ [     ] $)$ dollars

(3) Find the equation for the cost.

The price for $1 stamps

[     ] $+ 1.20 ($

The price for $1.20 stamps

[     ] $) =$

Total price

[     ] dollars

(4) How many stamps that cost $1 are there?

⟨Ans.⟩                    stamps

(5) How many stamps that cost $1.20 are there?

⟨Ans.⟩                    stamps

---

### Hint

Since there are two quantities to solve for, one quantity has to be represented as $x$ and the other by using $x$ in
the formula.

If the stamps that cost $1.20 are $x$ stamps, then the stamps that cost $1 are $(10 - x)$ stamps.

The equation is : $1.2x + 1(10 - x) = 11$

**1** When Jonathan bought 12 peaches and pears, at $1.50 per peach and    20 points
at $1.60 per pear, the total price was $18.50.
How many peaches and pears did he buy, respectively?
Fill in the blanks provided and solve for the answer.

When Jonathan bought $x$ peaches,

The number of pears            $\Rightarrow$    [          ] pieces

The price of peaches           $\Rightarrow$    [          ] dollars

The price of pears             $\Rightarrow$    $1.60\,($ [          ] $)$ dollars

The equation for the price,

| The price of peaches + The price of pears | | Total price |
|---|---|---|
| [                ] | = | [          ] |

Solve the above equation.                                    〈Ans.〉 _____ peaches

                                                             〈Ans.〉 _____ pears

**2** When Anna bought 10 cream puffs and ice creams, at $1.50 per        15 points
cream puff and $1.20 per ice cream, the total price was $13.80.
How many cream puffs and ice creams did she buy, respectively?

                                                             〈Ans.〉 _____ cream puffs

                                                             〈Ans.〉 _____ ice creams

**3** Brenden bought 20 bananas and apples, at $1.20 per banana and       15 points
$1.40 per apple. The total cost was $26.
How many bananas and apples did he buy, respectively?

                                                             〈Ans.〉 _____ bananas

                                                             〈Ans.〉 _____ apples

# Linear Equations 7
## (Pattern 5)

Level

Score

/100

■ The Answer Key is on page 91.

Date      /      /

Name

---

## Pattern 5 — Division

● There are 50 sheets of drawing paper.
Larry divides the drawing paper between himself
and his younger brother John.
And Larry has 14 more sheets.
How many sheets of drawing paper do Larry and his
brother have, respectively?

---

★ **TRY!** — **Fill in the blanks provided and solve for the answer.**

15 points per question

( 1 )   If John has $x$ sheets of drawing paper,
how many pieces of drawing paper does Larry have?

John's sheets + Difference in
numbers between Larry's and
John's

[＿＿＿＿] sheets

( 2 )   Create an equation to find the number of sheets of drawing paper.

[＿＿＿] sheets  +  ( [＿＿＿] ) sheets  =  [＿＿＿] sheets

Solve the above equation.

〈**Ans.**〉 ＿＿＿＿ sheets

( 3 )   John has [＿＿＿] sheets of drawing paper.

Therefore, Larry's drawing paper is

18  +  [＿＿＿] sheets  =  [＿＿＿] sheets

〈**Ans.**〉 Larry ＿＿＿＿ sheets

〈**Ans.**〉 John ＿＿＿＿ sheets

---

### Hint

After finding $x$,

make sure to check that all pieces are represented and do not forget to find the second quantity.

---

**1**  A piece of tape is 180 cm.
If Ruth divided the tape between herself and her younger sister Ann,
Ruth's tape will be 60 cm longer than Ann's.
How long is each sister's tape in centimeters?
Fill in the blanks provided and solve for the answer.

15 points

If the length of the Ann's tape is $x$ cm,
the length of Ruth's tape is,

Ruth´s tape

☐ cm

The equation for the length of the tape,

Ruth´s tape + Ann´s tape          Overall length

☐ cm  =  ☐ cm

Solve the above equation.

⟨**Ans.**⟩  Ruth _____ cm

⟨**Ans.**⟩  Ann _____ cm


**2**  There are 41 students in 6th grade Group A of a junior high school,
and the number of boys is 3 more than the number of girls.
How many boys and girls are there in the class, respectively?

20 points

⟨**Ans.**⟩  Boys _____ students

⟨**Ans.**⟩  Girls _____ students


**3**  A senior class at Smith High School has 84 students.
The number of boys in the class is 4 more than the number of girls.
How many boys and girls are there in the class, respectively?

20 points

⟨**Ans.**⟩  Boys _____ students

⟨**Ans.**⟩  Girls _____ students

# Linear Equations 8
## (Pattern 6)

**18**

Level ★★

Date / /

Name

Score /100

■ The Answer Key is on page 91.

## Pattern 6 — Increase and Decrease

●Brenda has 41 marbles and her younger
sister has 13 marbles.
Their mother asked Brenda to give her sister
some of her marbles, so that she has 10 more
marbles than her sister.
How many marbles should Brenda give to her
sister?

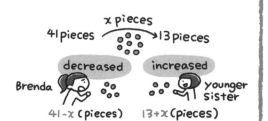

★ **TRY!** — Fill in the blanks provided and solve for the answer.

20 points per question

( 1 )  If Brenda passes $x$ marbles to her younger sister,

Brenda's marbles ⇒

Initial pieces − Pieces passed = pieces

The younger sister's marbles ⇒

Initial pieces + Pieces received = pieces

( 2 )  Find the equation for number of marbles.

Brenda's marbles pieces = Younger sister's marbles pieces + Difference of pieces between Brenda's and younger sister's pieces

Solve the above equation.

⟨Ans.⟩ _____ marbles

### Hint

Read the question sentences carefully and be careful not to make a mistake about which of the sisters has
10 more marbles.

**1** Justin has 26 cards and his younger brother has 11 cards.
Their mother wants Justin to hand some of his cards over to his younger
brother, so that Justin still has 5 more cards than his brother.
How many cards should Justin give to his brother?
Fill in the blanks provided and solve for the answer.

20 points

If Justin passes $x$ cards to his younger brother,

Justin's cards　　　　　　　　⇒

Justin's cards

[ ] cards

The younger brother's cards　　⇒

Younger brother's cards

[ ] cards

The equation for the number of cards,

Justin's cards

[ ] cards　=　

Younger brother's cards + Difference in the
number of cards between brothers

[ ] cards

Solve the above equation.　　　　　　　　　　〈Ans.〉　　　　　　cards

**2** Scott bought 14 cookies and Pamela bought 24 cookies.
Their teacher asked Pamela to give Scott some of her cookies,
so that he had 2 more cookies than her.
How many cookies should Pamela give Scott?

20 points

〈Ans.〉　　　　　　cookies

**3** David bought 28 cupcakes and Sydney bought 12 cupcakes.
David decided to give some of his cupcakes to Sydney so that she had
6 more cupcakes than him.
How many cupcakes should David give to Sydney?

20 points

〈Ans.〉　　　　　　cupcakes

**19**

Level

Score

Date / /

Name

/100

■ The Answer Key is on page 91.

## Pattern 7 — Excess and Shortage

● Sheets of colored paper are divided equally by some students.
If they divide the sheets by 4, there are 10 sheets left.
If they divide the sheets by 6, there is a shortage of 8 sheets.
How many students are there?

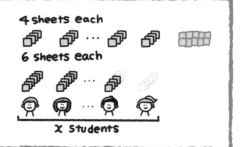

★ **TRY!** — Fill in the blanks provided and solve for the answer.

20 points per question

( 1 ) If the number of students is $x$, there are 10 sheets left, and therefore the total number of sheets is,

Number of sheets per student × Number of students

⬚ **+** Number of sheets left ⬚ sheets

( 2 ) If they divide the sheets by 6, there is a shortage of 8 sheets and so,

Number of sheets per student × Number of students

⬚ **−** Shortage of sheets ⬚ sheets

( 3 ) Since quantities of #1 and #2 above are equal, the equation about the sheets of colored paper is,

⬚ **+** ⬚ **=** ⬚ **−** ⬚ sheets

Solve the above equation.

⟨Ans.⟩ _____ students

### Hint

Since the number of colored paper does not change,

considering the excess or shortage,

Create an equation:

"When dividing the sheets by 4" = "When dividing the sheets by 6".

**1** Some students are trying to divide pencils evenly among them, so that each student has the same number of pencils.
If the students divide the pencils by 3, there are 15 pencils left.
If they divide the pencils by 5, there is a shortage of 1 pencil.
How many students are there?
Fill in the blanks provided and solve for the answer.

10 points

If the number of students is represented by $x$,
and the total number of pencils is divided by 3,
there are 15 pencils left,

| | pencils |
|---|---|

If the number of pencils is divided by 5, there is a shortage of 1 pencil,

| | pencils |
|---|---|

Create an equation for the number of pencils,

| |
|---|

Solve the above equation.

⟨Ans.⟩      students

**2** Mrs. Smith wants to give the same number of candies to several children.
When she gives each child 3 pieces, there is a shortage of 7 candies.
When she gives each child 2 pieces, there are 14 candies left.
How many children are there?

15 points

⟨Ans.⟩      children

**3** Joseph wants to share his gumballs with his friends.
If he gives each friend 4 gumballs, there are 12 gumballs left.
If he gives each friend 6 gumballs, there is a shortage of 4 gumballs.
How many friends are there?

15 points

⟨Ans.⟩      friends

**20**

# Linear Equations 10
## (Pattern 8)

Level ★★

Score

/100

Date / /

Name

■ The Answer Key is on page 91.

### Pattern 8 — Multiples of A

● A number is obtained by subtracting 4 units from 6 times a certain number $n$.
It is equal to a number obtained by adding 12 units to twice $n$.
Find the value of $n$.

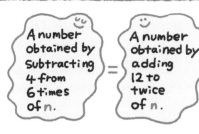

A number obtained by Subtracting 4 from 6 times of n. = A number obtained by adding 12 to twice of n.

★ **TRY!** — Fill in the blanks provided and solve for the answer.

20 points per question

( 1 ) When using a certain number $n$ and representing the relationship of numbers by two kinds of expressions,
A number obtained by subtracting 4 units from 6 times $n$ is,

6 times $n$ — Number subtracted

$$6\ \boxed{\phantom{xxx}} - \boxed{\phantom{xxx}}$$

A number obtained by adding 12 units to twice $n$ is,

2 times $n$ + Number added

$$2\ \boxed{\phantom{xxx}} + \boxed{\phantom{xxx}}$$

( 2 ) Since the two quantities of #1 above are equal, the equation for $n$ is,

$$6\ \boxed{\phantom{xxx}} - \boxed{\phantom{xxx}} = 2\ \boxed{\phantom{xxx}} + \boxed{\phantom{xxx}}$$

Solve the above equation.

⟨Ans.⟩ _____

**Hint**

Represent the relationship between numbers with two kinds of expressions and connect them with an equal sign.

## Key Point

"The number obtained by adding □ to ○ times A"
$$\Rightarrow \text{"A} \times \bigcirc + \square\text{"}$$
"The number obtained by subtracting ■ from △ times A"
$$\Rightarrow \text{"A} \times \triangle - \blacksquare\text{"}$$

**1** **A number obtained by adding 5 units to 3 times a certain number $x$ is equal to a number obtained by subtracting 9 units from 5 times $x$.**
**Find the value of $x$.**
**Fill in the blanks provided and solve for the answer.**

20 points

When a certain number $x$ is used to express the relationship of numbers
as two kinds of expressions,
the number obtained by adding 5 units to 3 times $x$ is,

The number obtained by subtracting 9 units from 5 times $x$ is,

The equation about numbers is,

Solve the above equation.

⟨Ans.⟩ _____

**2** **A volleyball player, Brandon, says,**
**"When I add 4 units to twice the number on my jersey, the value is same as when I subtract 8 units from 6 times my jersey's number."**
**What is the number on Brandon's jersey?**

20 points

⟨Ans.⟩ _____

**3** **James is a racecar driver.**
**The number on his racecar can be found when he adds 7 units to 3 times the number or when he subtracts 2 units from 6 times the number.**
**What is the number on James' racecar?**

20 points

⟨Ans.⟩ _____

■ The Answer Key is on page 92.

Date / /

Name

## Pattern 9 — Using the Nature of Division

● When 200 is divided by a certain number $x$,
the quotient is 15 and the remainder is 5.
Find the value of $x$.

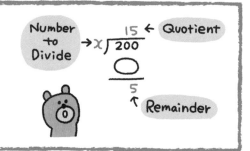

★ **TRY!** — Fill in the blanks provided and solve for the answer.    15 points per question

(1) First, consider a case where there is no remainder.
When a number is divided by a certain number $x$, the quotient is 15.

| Number to divide | | Quotient |
|---|---|---|
| | × | |

(2) When a number is divided by $x$, the quotient is 15 and the remainder is 5.

| Number to divide | | Quotient | | Remainder |
|---|---|---|---|---|
| | × | | + | |

Therefore,        15 [ ] + [ ]

(3) Create an equation if the number found in # 2 above is equal to 200,

15 [ ] + [ ] = [ ]

Solve the above equation.

⟨Ans.⟩ _____

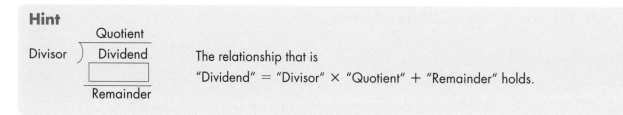

**Hint**

Quotient
Divisor ) Dividend
[ ]
Remainder

The relationship that is
"Dividend" = "Divisor" × "Quotient" + "Remainder" holds.

**1** If 8 times a certain number $x$ is divided by 9, the quotient is 5 and the remainder is 3.
Find the value of $x$.
Fill in the blanks provided and solve for the answer.

15 points

The product of 8 times $x$ is,

$$\boxed{\phantom{xxxx}}$$

Express the number whose quotient is 5 and the remainder is 3 as,

$$9 \quad \times \quad \boxed{\phantom{xxx}} \quad + \quad \boxed{\phantom{xxx}}$$

Therefore,

$$\boxed{\phantom{xxx}}$$

Create an equation,

$$\boxed{\phantom{xxxxx}}$$

Solve the above equation.          〈**Ans.**〉 _____

**2** If 12 times a certain number $x$ is divided by 7, the quotient is 6 and the remainder is 2 larger than the original number $x$.
Find the value of $x$.

20 points

〈**Ans.**〉 _____

**3** If 7 times a certain number $x$ is divided by 9, the quotient is 5 and the remainder is 3 smaller than the original number $x$.
Find the value of $x$.

20 points

〈**Ans.**〉 _____

# Linear Equations 12
## (Pattern 10)

Level
☆☆

Score

/ 100

Date / /

Name

■ The Answer Key is on page 92.

**Pattern 10 — The Two-Digit Natural Number**

● There is a two-digit natural number where the number in the ones place is 7.
When the number in the tens place and the number in the ones place are switched,
it becomes 9 units larger than the original natural number.
Find the original natural number.

replacement

★ **TRY!** — Fill in the blanks provided and solve for the answer.

10 points per question

( 1 ) If the tens place of the original natural number is defined as $x$,
the original natural number is,

Number of tens place    Number of ones place

10 ☐ + ☐

( 2 ) The natural number whose numbers in the tens place and ones place are switched is,

70 + ☐

( 3 ) Therefore, the equation is,

Replaced ones place        Original natural number

70 + ☐ = ( 10 ☐ + ☐ ) + 9

Solve the above equation.           ⟨Ans.⟩ _____

( 4 ) The number in the tens place of the original number is,

☐

Therefore, the original number is,

☐ × 6 + 7 = ☐

⟨Ans.⟩ _____

**Hint**

In the word problem of an equation, the answer is often represented by $x$, but in the above question, since the ones place is given as 7, let the tens place be $x$.
Try and clarify what is known and what is unknown in the question.

## Key Point

| | |
|---|---|
| If the tens place is a and the ones place is b, the two-digit natural number is, | $10a + b$ |
| If the tens place and the ones place are switched, the two-digit natural number is, | $10b + a$ |

**1** **There is a two-digit natural number where the ones place is 3.** 20 points
**If the tens place and ones place are switched,**
**it is 27 units smaller than the original natural number.**
**Fill in the blanks provided and solve for the answer.**

If the number in the tens place of the original natural number is defined as $x$,
the original natural number is,

[ ]

The natural number whose tens place and ones place are switched is,

[ ]

The equation is,

Replaced number          Original number − 27

[ ] = [ ]

Solve the above equation and find the original number.

⟨Ans.⟩ _____

**2** **There is a two-digit natural number where the tens place is 8.** 20 points
**If the tens place and ones place are switched,**
**it is 36 units smaller than the original natural number.**
**Find the original number.**

⟨Ans.⟩ _____

**3** **There is a two-digit natural number where the tens place is 5.** 20 points
**If the tens place and ones place are switched,**
**it is 18 units larger than the original natural number.**
**Find the original number.**

⟨Ans.⟩ _____

Date / /

Name

■ The Answer Key is on page 92.

---

**Pattern 11 — Age**

● Currently, Nicole's father is 46 years old and Nicole is 14 years old.
How many years does it take until Nicole's father's age is 3 times Nicole's age?

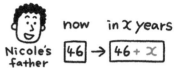

now    in $x$ years

Nicole's father    $46 \rightarrow 46 + x$

Nicole    $14 \rightarrow 14 + x$

---

★ **TRY!** — **Fill in the blanks provided and solve for the answer.**

20 points per question

(1)  Create an equation where 3 times the age in $x$ years.
Then, the ages of the two people in $x$ years are as follows.

Nicole's father age    $\Rightarrow$    46 + ☐ years old

Nicole's age    $\Rightarrow$    14 + ☐ years old

(2)  The equation about the ages of the two people in $x$ years is,

Nicole's father age
☐  = 3 ( ☐ )
Nicole's age

Solve the above equation.

⟨**Ans.**⟩ _____ years later

---

**Hint**

Since the two people will get older by the same number of years in $x$ years,

the age in $x$ years should be "$+ x$".

The age $x$ years before is "$- x$".

---

**1** Currently, Frank is 12 years old and his mother is 44 years old.    20 points
How many years later will his mother's age become 3 times Frank's age?
Fill in the blanks provided and solve for the answer.

Assuming that his mother's age become 3 times Frank's age $x$ years later,
the ages of the two people in $x$ years are,

     Frank's age          ⇒    [    ] years old

     Frank mother's age    ⇒    [    ] years old

The equation about the ages of the two people in $x$ years is,

$$\underset{\text{Frank mother's age}}{\boxed{\phantom{xxxxx}}} = 3\left(\underset{\text{Frank's age}}{\boxed{\phantom{xxxxx}}}\right)$$

Solve the above equation.

⟨Ans.⟩ _____ years later

**2** Beth is 7 years old and her elder sister Jo is 17.    20 points
How many years later will Jo's age be twice Beth's age?

⟨Ans.⟩ _____ years later

**3** Katherine is 15 years old and her dad is 45 years old now.    20 points

How many years ago was Katherine's age $\frac{1}{4}$ of her father's age?

⟨Ans.⟩ _____ years ago

# Linear Equations 14
## (Pattern 12)

24

Level ☆☆

Score /100

Date / /

Name

■ The Answer Key is on page 92.

---

### Pattern 12 — When Catching up

● Benjamin left home and headed to the station.
2 minutes later, his elder sister left from the same house and ran after him.
If his walking speed is 60 meters per minute (m/min),
and his sister's walking speed is 70 meters per minute (m/min),
how long will it take for her to catch up to Benjamin?

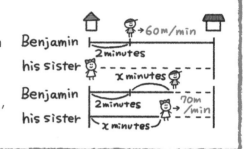

---

★ **TRY!** — Fill in the blanks provided and solve for the answer.

20 points per question

( 1 ) Suppose Benjamin's sister catches up with him in $x$ minutes.
The distance Benjamin walked is,

$$\boxed{\phantom{Speed}} \times (\ 2\ +\ \boxed{\phantom{Time}}\ )\ \text{(m)}$$

Speed        Time

The distance his elder sister walked is,

$$\boxed{\qquad x}\ \text{(m)}$$

Speed × Time

( 2 ) The equation for the distance Benjamin and his sister walked is,

$$\boxed{\phantom{x}} \times (\ 2\ +\ \boxed{\phantom{x}}\ ) = \boxed{\qquad x}\ \text{(m)}$$

Distance Benjamin walked                    Distance his sister walked

Solve the above equation.

〈**Ans.**〉　　　　　　　　minutes later

---

**Hint**

Since the distance is the same,

we can set up an equation;

"Distance Benjamin walked" = "Distance his sister walked"

**1** **Samantha departed from her house and walked to her uncle's house on foot.** 20 points
**12 minutes later, her elder brother noticed Samantha had forgotten**
**something and chased her by bike.**
**If Samantha walks at a speed of 80 m/min and the speed of her brother's bike**
**is 200 m/min, how many minutes does it take for her brother to catch up to her**
**after he leaves home?**
**Fill in the blanks provided and solve for the answer.**

If her brother catches up with Samantha in $x$ minutes after Samantha's departure,
The distance Samantha traveled is,

$$80 \left( \phantom{xxxxx} \right) \text{ (m)}$$

The distance the elder brother traveled is,

$$\boxed{\phantom{xxxxx}} \text{ (m)}$$

The equation for how far the 2 siblings traveled is,

Distance Samantha walked          Distance her brother walked

$$80 \left( \phantom{xxxx} \right) = \boxed{\phantom{xxxx}}$$

Solve the above equation.

⟨**Ans.**⟩ _____ minutes later

**2** **Christian left home for his aunt's house on foot.** 20 points
**15 minutes later, his elder sister found things Christian had forgotten and**
**chased him by bicycle.**
**If Christian's walking speed is 60 m/min and his sister's bicycle speed is**
**210 m/min, how many minutes later does his sister catch up with Christian?**

⟨**Ans.**⟩ _____ minutes later

**3** **Sophie left home on foot toward school.** 20 points
**6 minutes later, her brother found Sophie's homework on the kitchen table**
**and left the house to chase after her on his bike.**
**If Sophie's walking speed is 40 m/min and her brother's biking speed**
**is 100 m/min, how long will it take for him to catch up to Sophie?**

⟨**Ans.**⟩ _____ minutes later

Date / /

Name

■ The Answer Key is on page 92.

## Pattern 13 — Meeting Each Other by Going in the Opposite Direction

●There is a pedestrian path that is a 1,800 m long circle.
Jacqueline started walking at a pace of 70 m/min and
Ethan at a pace of 80 m/min.
They both departed from the same point at the same time,
but walking in the opposite direction.
How many minutes later do they meet for the first time
after the departure?

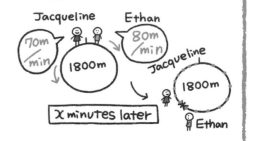

★ TRY! — Fill in the blanks provided and solve for the answer.    20 points per question

(1) Assuming that Jacqueline and Ethan will meet for the first time in $x$ minutes after departing,
The distance Jacqueline walked is,

Speed × Time

[          $x$ ] (m)

The distance Ethan walked is,

Speed × Time

[          $x$ ] (m)

(2) The equation for how far the two people walked is,

Distance Jacqueline walked        Distance Ethan walked        Distance per lap

[          ]  +  [          ]  =  [          ]

Solve the above equation.

⟨Ans.⟩ _____ minutes later

### Hint

When two people go in the opposite direction,

it can be considered that the sum of the distance advanced by the two people is equal to the distance along an arc per lap.

Therefore,

"Distance Ms. A walked" + "Distance Mr. B walked" = "Distance per lap"

## Key Point

When two people are going in the opposite direction on the circumference of a circle and encounter each other,

"Sum of distance traveled by two people" = "Distance per lap"

**1** There is a bicycle only road that is 2,400 m per lap.
Hannah started to cycle at 240 m/min and Austin at 160 m/min from the same spot on the bicycle path at the same time, but in opposite directions.
How many minutes later do they meet for the first time after the departure?
Fill in the blanks provided and solve for the answer.

*20 points*

Assuming the two people first meet $x$ minutes after departing,
The distance Hannah rode is,

$\boxed{\phantom{XXXXXXX}}$ (m)

The distance Austin rode is,

$\boxed{\phantom{XXXXXXX}}$ (m)

The equation for the distance the two people rode is,

$$\underset{\text{Distance of Hannah +}{\text{Distance of Austin}}}{\boxed{\phantom{XXXX}}} = \underset{\text{Distance per lap}}{\boxed{\phantom{XXXX}}}$$

Solve the above equation.

⟨Ans.⟩ _____ minutes later

**2** There is a bike only road that is 12 km per lap.
Doris's speed is 10 km/h, and Joe's speed is 14 km/h.
They departed from the same spot at the same time in the opposite direction.
How many hours later do they meet for the first time after the departure?

*20 points*

⟨Ans.⟩ _____ hours later

**3** The big park has a bike trail that is 40 miles per lap.
Geoff rides his bike at a speed of 13 miles per hour and Cindy rides her bike at a speed of 7 miles per hour.
They begin their ride at the same spot and at the same time in the opposite direction.
How many hours later do they meet for the first time after the departure?

*20 points*

⟨Ans.⟩ _____ hours later

Date / /

Name

■ The Answer Key is on page 92.

## Pattern 14 — Proceeding in the Same Direction and Catching up

● There is a promenade of 800 meters long per lap around a park.
Kathryn and Albert started to run in the same direction
simultaneously from the same spot.
Kathryn's running speed is 200 m/min and Albert's is 160 m/min.
How many minutes later does Kathryn catch up to Albert for the
first time after she begins running?

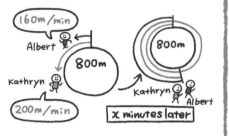

★ **TRY!** — **Fill in the blanks provided and solve for the answer.**

20 points per question

(1) Assuming that Kathryn catches up with Albert $x$ minutes later for the first time
after they began running,
The distance which Kathryn ran is,

Speed × Time

| $x$ | (m) |

The distance which Albert ran is,

Speed × Time

| $x$ | (m) |

(2) The equation for the distance which the two people ran is,

| Distance Kathryn ran | — | Distance Albert ran | = | Distance per lap |

Solve the above equation.

⟨**Ans.**⟩ _____ minutes later

### Hint

When two people go in the same direction,

it can be considered that the difference in distance which they advanced is equal to the one-way trip
per lap,

"Distance Kathryn ran" — "Distance Albert ran" = "Distance per lap"

**1** There is an athletic track that is 400 m long per lap.

20 points

Gloria walks at a speed of 70 m/min and Jesse jogs at a speed of 170 m/min.
They begin to move forward in the same direction from the same point at the same time.
How many minutes later does Gloria catch up with Jesse for the first time after beginning?
Fill in the blanks provided and solve for the answer.

Assuming Jesse first catches up with Gloria $x$ minutes after they begin to move forward,
The distance Gloria moved was,

$\boxed{\phantom{xxxxxxxxxx}}$ (m)

The distance Jesse moved was,

$\boxed{\phantom{xxxxxxxxxx}}$ (m)

The equation for the distance that they moved is

| Distance of Jesse —<br>Distance of Gloria | | Distance of track per lap |
|---|---|---|
| $\boxed{\phantom{xxxxxxxx}}$ | = | $\boxed{\phantom{xxxxxxxx}}$ |

Solve the above equation.

⟨Ans.⟩ _____ minutes later

**2** There is a 2,000 m long road around a lake.

20 points

Teresa was on foot and Willie rode a bicycle at a speed of 300 m/min.
They started to move forward in the same direction simultaneously from the same point.
When Willie first caught up with Teresa in 8 minutes, how fast was Teresa's walking speed in m/min?

⟨Ans.⟩ _____ m/min

**3** Thomas and Anthony are planning to hike a 5.1 km trail loop.

20 points

Thomas plans to walk and Anthony plans to ride his bicycle at a speed of 400 m/min.
They start hiking at the same time from the same starting place.
Anthony first catches up to Thomas after 15 minutes,
how fast is Thomas's walking speed in m/min?

⟨Ans.⟩ _____ m/min

# Linear Equations 17
## (Pattern 15)

**27**

Date    /    /

Name

Level
★★

Score
/100

■ The Answer Key is on page 92.

---

## Pattern 15 — Thinking About the Difference in Time Required

● A person climbs a mountain road from the base to the summit at a speed of 45 m/min and descends the same road from top to bottom at a speed of 60 m/min.
When comparing the length of time, there is a 20 minute difference between the climbing and descending times.
How many meters is it from the base to the summit of the mountain?

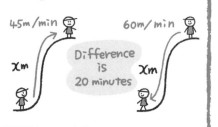

---

★ **TRY!** — Fill in the blanks provided and solve for the answer.

20 points per question

（1） Suppose the distance from the base to the summit of the mountain is $x$ meters, after a person begins climbing,
the time required from the base to the summit is,

Distance
─────── (min)
45

The time required from the summit to the base is,

Distance
─────── (min)
60

（2） The equation for the required time is,

From the base to the summit    From the summit to the base

☐ − ☐ = 20

Solve the above equation.

〈Ans.〉 _____ m

---

### Hint

45 m/min means to advance 45 meters per minute.

60 m/min means to advance 60 meters per minute.

The longer length of time is the speed of 45 m/min, so the difference in time is,

$$\frac{x}{45} - \frac{x}{60}$$

54    © Kumon Publishing Co., Ltd.

**Key Point**

Between cases A and B, when case A is longer by $x$ minutes,
"Time required with A" − "Time required with B" $= x$.

**1** To travel from City A to City B at a speed of 4 km/h and to go back from City B to City A at a speed of 5 km/h creates a 1 hour difference in the length of time it takes to travel between the cities.
How many kilometers is it from City A to City B?
Fill in the blanks provided and solve for the answer.

20 points

Assuming that distance from City A to City B is $x$ km,
The length of time from City A to City B is,

☐ hours

The length of time from City B to City A is

☐ hours

The equation for the length of time required is,

Time from City A to City B −
Time from City B to City A

☐ $= 1$

Solve the above equation.

⟨Ans.⟩ _____ km

**2** From City A to City B, Andrew could ride his bicycle at 14 km/h or walk the same road at a rate of 6 km/h.
There is a 2 hours time difference between Andrew riding or walking.
How many kilometers is it from City A to City B?

20 points

⟨Ans.⟩ _____ km

**3** Arya wants to visit her sister's house. She can ride her bike at 12 km/h or she can walk at a speed of 6 km/h.
There is a $2\frac{1}{3}$ hour difference in the time it takes her to travel from her house to her sister's house when walking or cycling.
How many kilometers is it from Arya's house to her sister's house?

20 points

⟨Ans.⟩ _____ km

# Linear Equations 18
## (Pattern 16)

Score /100

Date / /

Name

■ The Answer Key is on page 93.

---

## Pattern 16 — Proportional Expression to Consider the Ratio of Quantities

● When Janice makes a dressing, she usually mixes 70 mℓ of vinegar and 140 mℓ of olive oil.
When there is only 100 mℓ of olive oil, how many milliliters of vinegar does she need to make the dressing taste the same?

Usual taste — vinegar : Olive Oil — 70mL : 140mL

Same taste — XmL : 100mL

---

★ **TRY!** — Fill in the blanks provided and solve for the answer.

20 points per question

(1)  Suppose the amount of vinegar needed is $x$ mℓ,
The ratio for the usual amount of vinegar to olive oil is,

Olive oil

70 : ☐

The ratio of the amount of vinegar to the amount of olive oil when there is 100 mℓ of olive oil,

Vinegar          Olive oil

☐ : ☐

(2)  The proportional expression of vinegar to olive oil is,

Ratio when olive oil is 100 mℓ          Ratio between usual vinegar and olive oil

☐ = ☐

Solve the above proportional expression.

⟨Ans.⟩ _____ mℓ

---

**Hint**

When making a proportional expression,
try not to mistake the order of the ratios.

Nature of the proportional expression     $a : b = c : d \Rightarrow ad = bc$

**1** **When Marie makes pancakes, she mixes at a ratio of 150 g of flour**                   20 points
**to 60 g of butter.**
**She prepared 400 g of flour to make pancakes with the same taste.**
**How many grams of butter does she need?**
**Fill in the blanks provided and solve for the answer.**

Suppose she needs $x$ g of butter.
The ratio of the usual amount of butter and flour is,

$$60 \quad : \quad \boxed{\phantom{xxxx}}$$

When she prepared 400 g of flour, the ratio of the amount of butter and flour is,

$$\boxed{\phantom{xxxx}} \quad : \quad \boxed{\phantom{xxxx}}$$

The proportional expression for the amount of butter to flour is,

$$\boxed{\phantom{xxxx}} = \boxed{\phantom{xxxx}}$$

Solve the above proportional expression.

⟨Ans.⟩ _____ g

**2** **Julia mixes at a ratio of 200 g of wheat flour to 80 g of butter when**                   20 points
**making sweets.**
**To make the same tasting sweets, 350 g of flour was prepared.**
**How many grams of butter does she need?**

⟨Ans.⟩ _____ g

**3** **When Kyle makes chocolate fudge,**                   20 points
**he mixes 400 g of cocoa powder to 260 ounces of milk.**
**To make the fudge taste the same with 700 g of cocoa powder,**
**how many ounces of milk does he need?**

⟨Ans.⟩ _____ ounces

■ The Answer Key is on page 93.

## Pattern 17 — Proportional Expression to Consider the Ratio to the Total

● Grace wants to divide 140 sheets of colored paper between herself and her younger sister.
Grace wants to divide the sheets at a ratio of 4 : 3, so that Grace has more sheets.
How many sheets of colored paper will they each have?

140 Sheets

4 : 3

Grace     her younger sister

## ★ TRY! — Fill in the blanks provided and solve for the answer.

15 points per question

(1) Define Grace's number of sheets as $x$.
Since the ratio of Grace's and her younger sister's sheets is 4 : 3,
the ratio of the number for Grace to the total number of sheets is,

Total number

☐ : 4

(2) The number of the total sheets is,

☐ sheets

The number of Grace's sheets is,

☐ sheets

(3) The proportional expression is,

Total number : Grace's number

140 : ☐ = ☐

Solve the above proportional expression.

〈Ans.〉 _____ sheets

### Hint

When the total amount is known,

It is easy to understand using the ratio to the total

(the total is $a + b$ when $a$ and $b$ is $a : b$).

In the question above, use the ratio;

"The total number of sheets : The number of Grace's sheets" = $(4 + 3)$ : 4.

**1** Bryan wants to divide 36 caramels between himself and his younger brother at a ratio of 5 : 4.
How many caramels should Bryan's brother have?
Fill in the blanks provided and solve for the answer.

15 points

Assume the number for his brother is $x$ caramels.
The ratio of the number for his brother to the total number is,

$$\boxed{\phantom{xxxx}} : \quad 4$$

The total number is,

$$\boxed{\phantom{xxxx}} \text{ caramels}$$

The number of caramels his brother has

$$\boxed{\phantom{xxxx}} \text{ caramels}$$

The proportional expression is,

$$\boxed{\phantom{xxxx}} = \boxed{\phantom{xxxx}}$$

Solve the above proportional expression.

⟨Ans.⟩ _____ caramels

**2** Judy has 84 marbles. She wants to share the marbles with her younger sister at a ratio of 7 : 5.
How many marbles will Judy's sister receive?

20 points

⟨Ans.⟩ _____ marbles

**3** Neil has a package of 56 crayons. He wants to divide the crayons between himself and his younger brother at a ratio of 5 : 3.
How many crayons will Neil's brother receive?

20 points

⟨Ans.⟩ _____ crayons

■ The Answer Key is on page 93.

**1** Make an equation and find the number that applies to $x$ in the following sentences.

10 points per question

(1) The number obtained by adding 16 units to twice $x$ is equal to the number obtained by subtracting $x$ from 85.

⟨Ans.⟩ _____

(2) The number obtained by adding 5 units to 3 times $x$ is 1 unit larger than twice $x$ minus 2.

⟨Ans.⟩ _____

**2** Louis has $12 and his brother Russell has $4.
Louis bought 10 notebooks and Russell bought 3 of the same notebooks at the same price.
Louis's remaining money was twice as much as Russell's remaining money.
What is the price of 1 notebook?

10 points

⟨Ans.⟩ $ _____

**3** It took $2.10 to buy 1 notebook and 1 pencil.
Also, it took $5.10 to buy 5 of the same pencils and 1 of the same notebook.
Find the price for 1 notebook and 1 pencil.

10 points

⟨Ans.⟩ Notebook $ _____

⟨Ans.⟩ Pencil $ _____

**4** Gabriel distributes oranges to children gathered at the birthday party.
When distributing 6 oranges to each child, 23 oranges are left, and when 9 oranges are distributed to each child, 5 oranges are left.
How many children are there?

10 points

⟨Ans.⟩ _____ children

**5** Brittany went back and forth between point A and point B, going she walked at a speed of 6 km/h and on her return she walked at a speed of 4 km/h. And she took 7.5 hours. Find the distance between point A and point B.

⟨Ans.⟩ _____ km

**6** The population of City A this year has increased by 8% from last year to 16,200 people. Find the population of City A last year.

⟨Ans.⟩ _____ people

**7** In a 3% salt solution, 3% of the total weight of the saline solution is the weight of salt. Find how many grams of salt and water are contained in 1 kg of a 3% salt solution.

⟨Ans.⟩ Salt _____ g

⟨Ans.⟩ Water _____ g

**8** Randy bought a book with a quarter of his money.

And when he bought candy using $\frac{2}{5}$ of the remaining money, $5.40 remained.

How much money did he have initially?

⟨Ans.⟩ $ _____

Date      /      /

Name

■ The Answer Key is on page 93.

**1**   Diana has $2.                                                                          10 points
When she bought 2 notebooks of the same price, $0.20 remained.
What is the price of 1 notebook?

⟨Ans.⟩  $ _____

**2**   Abigail has $6, and her sister Natalie has $1.                             10 points
Then, they each got the same amount of money from their mother.
Abigail's total amount of money became twice as much as Natalie's money.
How much money did Abigail and Natalie receive from their mother?

⟨Ans.⟩  $ _____

**3**   It took $2 to buy 1 orange and 1 apple.                                       10 points
It took $12.40 to buy 5 of the same oranges and 7 of the same apples.
Find the price for 1 orange and 1 apple.

⟨Ans.⟩  Orange   $ _____

⟨Ans.⟩  Apple     $ _____

**4**   Philip's teacher distributed colored pencils to all the students in the class.   10 points
If he distributed 5 to each student 45 pencils remained.
If he distributed 8 to each student 6 pencils remained.
Find the number of students and the number of colored pencils.

⟨Ans.⟩ _____ students

⟨Ans.⟩ _____ colored pencils

**5** From the point P, Gregory chased Jane walking 600 meters, Jane proceeds at a speed of 80 m/min and Gregory proceeds at a speed of 200 m/min.
How many minutes later will Gregory catch up with Jane after leaving the point P?

15 points

⟨Ans.⟩ _____ minutes later

**6** When Christine sold a certain item for $7.20, she got a profit of 20% of the cost.
Find the original cost of this item.

15 points

⟨Ans.⟩ $ _____

**7** There is 500 g of a 10% salt solution.
How many grams of salt is dissolved in this saline solution?
Next, if water is added to dilute this saline solution to 4%, how many grams of water were added?

15 points

⟨Ans.⟩ Salt _____ g

⟨Ans.⟩ Water _____ g

**8** Virginia bought a book with a quarter of her money.
Continuing, she bought candy for $5.
Then, Virginia's money became $0.40 more than half of the money she had at the beginning.
How much money did she initially have?

15 points

⟨Ans.⟩ $ _____

**Proportions & Inverse Proportions 1**

Level ☆

Score

/100

Date / /

Name

■ The Answer Key is on page 94.

---

## ★ Check the basics before continuing – practice based questions!

Find words and numbers that apply to ①, ② and ③.

### (1) Function

There are two variables $x$ and $y$ which change together.
If the value of $x$ and $y$ is determined as having a single value,
it can be said that, "$y$ is a function of $x$".

#### ★ TRY!

6 points per question

The price when buying $x$ pencils at $0.60 is $y$ dollars.
At this time, if the value of $x$ had a single value ① is determined
accordingly, $y$ is ② of $x$.

⟨Ans.⟩ ① = _____

⟨Ans.⟩ ② = _____

### (2) Proportion

When $y$ is a function of $x$ and is expressed by $y = kx$ ($k$ is a constant),
it can be said that, "$y$ is proportional to $x$" and "$k$ is called a constant of proportionality".

#### ★ TRY!

7 points per question

James bought $x$ pieces of construction paper at $0.50 per sheet and the total price was $y$,
it is expressed as $y = $ ① and $y$ is ② to $x$.
Also, the constant of proportionality is ③.

⟨Ans.⟩ ① = _____

⟨Ans.⟩ ② = _____

⟨Ans.⟩ ③ = _____

### (3) Coordinate

In the figure on the right,
the set of numbers $(2, 3)$ represent point A and can be called the "coordinates"
of point A, 2 is the "$x$ coordinate", and 3 is the "$y$ coordinate".

#### ★ TRY!

7 points per question

In the figure on the right,
the coordinates of point B are ( ① , ② )

⟨Ans.⟩ ① = _____

⟨Ans.⟩ ② = _____

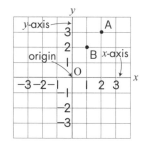

© *Kumon Publishing Co., Ltd.*

## (4) Proportional Graph

The graph of $y = kx$
($k$ is a constant) is a straight line
passing through the origin.

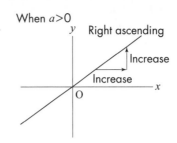

When $a > 0$

Right ascending

Increase

Increase

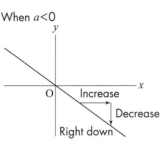

When $a < 0$

Increase

Decrease

Right down

### ★ TRY!

7 points per question

The graph of $y = 4x$ is ① that rises to the right passing through ② .

⟨**Ans.**⟩ ①= _____

⟨**Ans.**⟩ ②= _____

## (5) Inverse proportion

When $y$ is a function of $x$ and it is represented by the expression of
$y = \dfrac{k}{x}$ ($k$ is a constant),
it can be said that "$y$ is inversely proportional to $x$",
and $k$ is called a "constant of proportionality".

### ★ TRY!

7 points per question

Assuming that the length of a rectangle with an area of 36 cm³ is $x$ cm
and the width is $y$ cm, it is expressed $y = \dfrac{①}{x}$ and $y$ is ② to $x$.
Also, the constant of proportionality is ③ .

⟨**Ans.**⟩ ①= _____

⟨**Ans.**⟩ ②= _____

⟨**Ans.**⟩ ③= _____

## (6) Inverse Proportional Graph

The graph of $y = \dfrac{k}{x}$ ($k$ is a constant) is a smooth curve called
a "hyperbolic curve" as shown in the figure below.

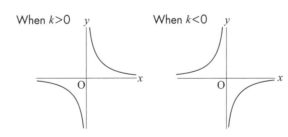

When $k > 0$    When $k < 0$

### ★ TRY!

6 points per question

The graph of $y = \dfrac{6}{x}$ is ① passing through coordinates (2, ②), ($-3$, ③) etc.

⟨**Ans.**⟩ ①= _____

⟨**Ans.**⟩ ②= _____

⟨**Ans.**⟩ ③= _____

# 33 Proportions & Inverse Proportions 2 (Pattern 1)

Level

Score

Date / /

Name

/100

■ The Answer Key is on page 94.

## Pattern 1 — The Number of Sheets and Thickness

● When measured, the thickness of 60 sheets of paper was 0.6 cm.
How thick will 1,300 sheets of the same paper be in centimeters?

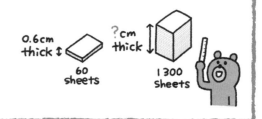

★ **TRY!** — Fill in the blanks provided and solve for the answer.

15 points per question

(1) Assume that the thickness of $x$ sheets of paper is $y$ cm.
Then, $y$ is proportional to $x$, using $k$.

$$y = \boxed{\phantom{xxx}}$$

(2) When $x = 60$, $y = 0.6$,

$$k = \frac{y}{x} = \frac{0.6}{60} = \frac{\boxed{\phantom{xx}}}{\boxed{\phantom{xx}}}$$

The expression is,

$$y = \boxed{\phantom{xx}} x$$

(3) The thickness of 1,300 sheets of paper is,

$$\frac{1}{100} \times 1300 = \boxed{\phantom{xx}} \text{ cm}$$

⟨Ans.⟩ _____ cm

### Hint

When the value of $x$ is multiplied by ☐
and the value of $y$ also becomes multiplied by ☐,
the relationship is proportional, so you can assume that $y = kx$.

**1** There are 300 of the same type of nails, altogether they weigh 540 g.
What is the weight of 45 nails?
Fill in the blanks provided and solve for the answer.

15 points

If the weight of $x$ nails is $y$ g,
since $y$ is proportional to $x$, the expression by using $k$ is represented as, ☐

When $x = $ ☐ , $y = $ ☐

From $k = \dfrac{\boxed{\phantom{xx}}}{\boxed{\phantom{xx}}} = $ ☐ $y = $ ☐ $x$

Solve to find the weight of 45 nails. ⟨**Ans.**⟩ _____ g

**2** There is a wire which is rolled up a few times and bundled.
The weight of 5 m of this wire is 90 g.
What is the weight of 3 m of the same wire?

20 points

⟨**Ans.**⟩ _____ g

**3** A piece of rope is coiled up into a bundle.
The weight of 8 ft of the rope coil is 120 pounds.
What is the weight of 5 ft of the same rope?

20 points

⟨**Ans.**⟩ _____ pounds

Date / /

Name

■ The Answer Key is on page 94.

## Pattern 2 — Weight and Area

● Tom made the shapes as shown on the right with an aluminum plate with a constant thickness. By the way, (a) is a square of which one side is 15 cm.
When the weight of board (a) is 40 g and the weight of the board (b) is 56 g, how many square centimeters is area of board (b)?

(a)
15 cm
15 cm  40 g

(b)
56 g
? cm²

★ **TRY!** — Fill in the blanks provided and solve for the answer.

15 points per question

(1) Assume the area of the aluminum plate of $x$ g is $y$ cm²,
Since $y$ is proportional to $x$,
the expression by using $k$ is represented as,

$y = \boxed{\phantom{xxx}}$

(2) In the case of (a),

$y = 15 \times 15 = \boxed{\phantom{xxx}}$

From #1 above when $x = 40$,

$k = \dfrac{225}{40} = \boxed{\phantom{xxx}}$

The expression is,

$y = \boxed{\phantom{xxx}} x$

(3) From #2 above, the area of board (b) is,

$\dfrac{45}{8} \times 56 = \boxed{\phantom{xxx}}$ cm²

⟨Ans.⟩ _____ cm²

### Hint

When $y$ is proportional to $x$, express $y = kx$
and substitute the values of $x$ and $y$ to find $k$.

**1** The shapes in the figure below were created with cardboard. **15 points**
When cardboard (a) weighs 24 g and cardboard (b) weighs 45 g,
how many square centimeters is the area of cardboard (b)?
Fill in the blanks provided and solve for the answer.

Assume the area of a piece of cardboard of $x$ g is $y$. Since $y$ is proportional to $x$, the expression by using $k$ is represented as

In the case of (a)

(a)

$$x = \boxed{\phantom{xx}} \qquad y = 14 \times 16 = \boxed{\phantom{xx}}$$

Therefore

(b)

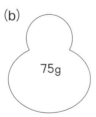

$$k = \frac{\boxed{\phantom{xx}}}{\boxed{\phantom{xx}}} = \boxed{\phantom{xx}} \qquad y = \boxed{\phantom{xx}}\ x$$

Find the area of cardboard (b).    〈Ans.〉 _____ cm²

**2** The shapes in the figure below were created with cardboard. **20 points**
When cardboard (a) weighs 48 g and cardboard (b) weighs 75 g,
what is the area of cardboard (b)?

(a)                (b)

〈Ans.〉 _____ cm²

**3** The shapes in the figure below were made with cardboard. **20 points**
When cardboard (a) weighs 36 g and cardboard (b) weighs 81 g,
what is the area of cardboard (b)?

(a)                (b)

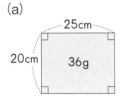

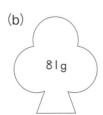

〈Ans.〉 _____ cm²

## 35

# Proportions & Inverse Proportions 4 (Pattern 3)

Level ☆☆

Score

Date / /

Name

/100

■The Answer Key is on page 94.

## Pattern 3 — Time and Distance

●It took 20 minutes to walk to the library which is located 1,600 m away from Bruce's house.
This time, assume Bruce is $y$ m away from the house $x$ minutes after leaving home.
Represent $y$ as the expression of $x$.

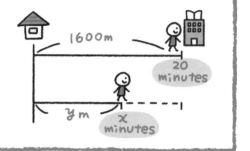

★ **TRY!** — Fill in the blanks provided and solve for the answer.

10 points per question

(1) The expression to find the speed is,

$\boxed{\phantom{XXXXX}}$ ÷ Time

(2) From #1 above, when walking 1,600 m in 20 minutes, the speed is,

1600 ÷ 20 = $\boxed{\phantom{XXXXX}}$ m/min.

(3) The expression to find the distance is,

$\boxed{\phantom{XXXXX}}$ × Time

(4) Since $y$ is proportional to $x$,

the distance $y$ when walking $x$ minutes at the speed of $\boxed{\phantom{XXXXX}}$ m/min is,

$y = \boxed{\phantom{XXXXX}} x$

⟨Ans.⟩ _____

### Hint

Sometimes, $x$ and $y$ cannot be defined as any values.

In the question above, to define the range of possible values of $x$ and $y$,

write "the domain of change" as follows:

$0 \leq x \leq 20$ and $0 \leq y \leq 1600$

**1** When Jordan walked up to the summit 1,200 m from the foot,  20 points
it took 24 minutes.
At this time, assume Jordan is at the point of $y$ m from the foot $x$ minutes
after leaving the foot and represent $y$ as the expression of $x$.
Fill in the blanks provided and solve for the answer.

The speed when walking 24 minutes for the distance of 1,200 m is,

Distance        Time

$$\boxed{\phantom{xxx}} \div \boxed{\phantom{xxx}} = \boxed{\phantom{xxx}} \text{ m/min.}$$

Since $y$ is proportional to $x$, the distance of $y$ m

when walking for $x$ minutes with the speed of $\boxed{\phantom{xxx}}$ m/min.

$$y = \boxed{\phantom{xxx}} x$$

⟨Ans.⟩ _____

**2** When Noah walked down to the foot of 1,200 m from the summit,  20 points
it took 15 minutes.
At this time, assume Noah is $y$ m from the summit $x$ minutes after leaving
the summit.
Represent $y$ as an expression of $x$.

⟨Ans.⟩ _____

**3** It took Dylan 3 hours to walk to B Town, which is 12 km away from A Town.  20 points
2 hours after Dylan departed from A Town, how many kilometers has he
walked from A Town?

⟨Ans.⟩ _____ km

■ The Answer Key is on page 94.

### Pattern 4 — Weight and Distance from a Fulcrum

● When a mobile is balanced, the value of "weight" × "distance from the fulcrum" becomes equal on the left and right sides. In the figure on the right, a weight of $x$ g is hung at the point of $y$ cm from the fulcrum, to balance the left and the right sides. At this time, assume $y$ is represented by the expression of $x$.

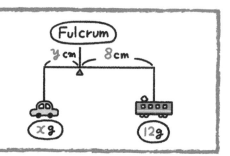

★ **TRY!** — Fill in the blanks provided and solve for the answer.          20 points per question

(1) When you hang a weight of $x$ g to the point of $y$ cm from the fulcrum,

(weight) × (distance from the fulcrum) =  ☐ × ☐

(2) When you hang a 12 g weight 8 cm from the fulcrum,

(weight) × (distance from the fulcrum) =  ☐ × ☐

(3) From the expressions of #1 and #2 above,

$$xy = \boxed{\phantom{xxxx}}$$

$$y = \boxed{\phantom{xxxx}}$$

⟨Ans.⟩ _____

---

**Hint**

When the mobile is balanced,

the value of "weight" × "distance from the fulcrum"

becomes equal on the left and right of the mobile,

so the expression of

"weight on the left side of the mobile × distance" =

"weight on the right side of the mobile × distance" holds.

## Key Point

Since the quantity of the weight and the distance from the fulcrum
are inversely proportional to each other,
assume $xy = k$ and find the constant of proportional $k$.

---

**1** When the mobile is balanced, the value of "the weight" ×
"distance from the fulcrum" becomes equal on the left and right sides.
In the figure below, when you hang the weight of $x$ g at the point of $y$ cm
from the fulcrum, the left and right sides balance.
At this time, assume $y$ is represented by the expression of $x$.
Fill in the blanks provided and solve for the answer.

20 points

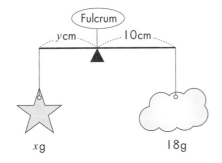

Since the left and right sides of the mobile are balanced,
the value of "weight" × "distance from the fulcrum" is
equal when lowering the weight of $x$ g at the point of $y$ cm
from the fulcrum and when lowering the weight of 18 g at
the point of 10 cm from the fulcrum.
Therefore,

$$x \times \boxed{\phantom{xx}} = \boxed{\phantom{xx}} \times \boxed{\phantom{xx}}$$

Assume $y$ is represented by the expression of $x$.

⟨Ans.⟩ _____

---

**2**

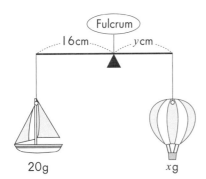

20 points

When the mobile is balanced, the value of
"weight" × "distance from the fulcrum"
becomes equal on the left and right sides.
In the figure on the left, when you hang the
weight of $x$ g at the point of $y$ cm from the
fulcrum, the left and right balanced.
When the weight on the right side is 8 g, how
many centimeters from the fulcrum are the left
and right balanced?

⟨Ans.⟩ _____ cm

# 37

## Proportions & Inverse Proportions 6 (Pattern 5)

Level

Score
/ 100

Date / /

Name

■ The Answer Key is on page 95.

**Pattern 5 — The Number of Rows and Chairs**

● 180 chairs were arranged in a gymnasium, so that the number of chairs in each row was the same as the other rows.

Assuming that when the number of rows is $x$ rows, the number of chairs arranged in 1 row is $y$ chairs, and $y$ is represented by the expression of $x$.

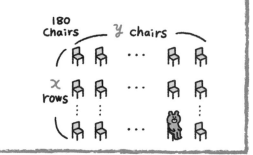

★ **TRY!** — Fill in the blanks provided and solve for the answer.

15 points per question

( 1 ) The expression to find the number of chairs to be arranged is,

Math symbol

(The number of rows) ☐ (The number of chairs arranged in 1 row)

( 2 ) Since the number of chairs is ☐ chairs altogether,

from #1 above, ☐ × ☐ = 180

( 3 ) From #2 above, if $y$ is represented by the expression of $x$,

$$y = \dfrac{\boxed{\phantom{xx}}}{\boxed{\phantom{xx}}}$$

⟨Ans.⟩

**Hint**

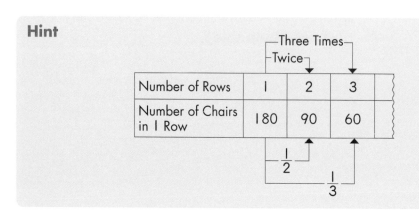

| | Three Times | | |
| Twice | | |
| Number of Rows | 1 | 2 | 3 |
| Number of Chairs in 1 Row | 180 | 90 | 60 |

## Key Point

When the total number does not change, it is considered that
the number of rows and the number of things arranged in 1 row
are inversely proportional.

**1**  **120 tulip bulbs were planted in a flowerbed, so that the number of tulip bulbs in 1 row was the same in each row.**
**If there are 15 rows, how many tulip bulbs are planted in 1 row?**
**Fill in the blanks provided and solve for the answer.**

15 points

Math symbol

(The number of rows) $\boxed{\phantom{x}}$ (The number of bulbs planted in 1 row) $= 120$

Therefore, if the number of rows is $x$ rows and the number of bulbs planted in 1 row is $y$,

The number of rows          The number in 1 row

$$x \quad \times \quad \boxed{\phantom{xxx}} \quad = \quad 120$$

If $y$ is represented by the expression of $x$,

$$y = \boxed{\phantom{xxx}}$$

If $x$ is 15, the value of $y$ is,

$$y = \frac{\boxed{\phantom{xxx}}}{15} = \boxed{\phantom{xxx}}$$

⟨Ans.⟩ _____ tulip bulbs

**2**  **270 tulip bulbs are planted in a flowerbed, so that the number of tulip bulbs planted in 1 row is the same in each row.**
**If there are 18 rows, how many tulip bulbs are planted in 1 row?**

20 points

⟨Ans.⟩ _____ tulip bulbs

**3**  **560 rose bushes are planted in a garden, so that the number of rose bushes planted in 1 row is the same in each row.**
**If there are 14 rows, how many rose bushes are planted in each row?**

20 points

⟨Ans.⟩ _____ rose bushes

# 38
## Proportions & Inverse Proportions 7 (Pattern 6)

Level ☆☆

Date  /  /.

Name

Score

/100

■The Answer Key is on page 95.

## Pattern 6 — The Amount of Water Added and the Time Required

●When water is added to a water tank at a rate of 6 ℓ per minute, it becomes full in 8 minutes.
Assuming water is added to this water tank at a rate of $x$ ℓ per minute and it is filled with water in $y$ minutes.
Represent $y$ as an expression of $x$.

## ★ TRY! — Fill in the blanks provided and solve for the answer.

15 points per question

(1)  The expression for finding the amount of water added is,

(The amount of water to be added per minute)  × (          )

(2)  Since the amount of water when it is full is [        ] ℓ,          [    ] × [    ] = 48

To add water for 8 minutes at the rate of 6 ℓ per minute

6 × 8

(3)  If $y$ is represented as the expression of $x$ by #2 above,

$$y = \frac{\boxed{\phantom{xx}}}{\boxed{\phantom{xx}}}$$

⟨Ans.⟩ _____

### Hint

When the water tank becomes full,

The amount of water put in $y$ minutes at the rate of $x$ ℓ per minute  =  The amount of water put in 8 minutes at the rate of 6 ℓ per minute

The expression above holds.

## Key Point

When the size of a water tank does not change, it is considered that
the time it takes to put in water and the amount of water put per minute
are an inverse proportion.

**1**  **When water is added at a rate of 7 ℓ per minute, a water tank becomes**    15 points
**full in 9 minutes.**
**How many minutes does it take to fill the water tank**
**if water is added at a rate of 3 ℓ per minute?**
**Fill in the blanks provided and solve for the answer.**

(The amount of water added per minute) $\times$ [            ] = (The amount of water added)

Therefore, if water is added at the rate of $x$ ℓ per minute and it becomes full in $y$ minutes,

Amount of water added per minute    Time (Minutes)    $7 \times 9$

$$x \quad \times \quad [\quad\quad] \quad = \quad 63$$

If $y$ is represented by the expression of $x$,

$$y = [\quad\quad]$$

If $x$ is 3, the value of $y$ is,

$$y = \frac{[\quad\quad]}{3} = [\quad\quad]$$

⟨Ans.⟩ _____ minutes

**2**  **When water is added at a rate of 6 ℓ per minute, a water tank becomes full**    20 points
**in 10 minutes.**
**How many minutes does it take to fill the water tank if water is added at a**
**rate of 5 ℓ per minute?**

⟨Ans.⟩ _____ minutes

**3**  **When water is added to a swimming pool at a rate of 36 gallons per**    20 points
**minute, the swimming pool becomes full in 68 minutes.**
**How many minutes does it take to fill the swimming pool if water is added**
**at a rate of 24 gallons per minute?**

⟨Ans.⟩ _____ minutes

## Proportions & Inverse Proportions Review 1

Date  /  /

Name

Score  /100

■ The Answer Key is on page 95.

**1** There is a car that runs 60 km on 5 ℓ of gasoline.
How many kilometers can that car run on 12 ℓ of gasoline?

5 points per question

（1） Find the answer considering how many times 12 ℓ is 5 ℓ of gasoline.

$$12 \div 5 = \frac{12}{5}$$

$$60 \times \frac{12}{5} =$$

〈Ans.〉 _____ km

（2） Find the answer considering the distance the car runs on 1 ℓ of gasoline.

$$60 \div 5 =$$

〈Ans.〉 _____ km

**2** The weight of a bar of iron that is 8 cm³ is 72 g.
What is the weight of the same bar of iron if it is 24 cm³?

10 points

〈Ans.〉 _____ g

**3** Water runs out of a tap at a rate of 45 ℓ in 5 minutes.
How many liters of water will run out of the tap in 12 minutes?

10 points

〈Ans.〉 _____ ℓ

**4** A piece of wire 6 m long weighs 85 g.
What is the weight of 24 m of this wire?

10 points

〈Ans.〉 _____ g

**5** The weight of 50 nails is 20 g.
What is the weight of 70 nails?

10 points per question

( 1 ) What is the weight of one of these nails?

⟨Ans.⟩ _____ g

( 2 ) Find the answer considering the weight of the nails is proportional to the number.

⟨Ans.⟩ _____ g

( 3 ) What is the weight of 110 nails with the answer of #1?

⟨Ans.⟩ _____ g

**6** Paper is stacked in a bundle.
The whole thickness of the paper bundle is 5 cm and the weight is 528 g.

10 points per question

( 1 ) When the thickness of the paper bundle is 1 cm, the number of sheets of paper is 88.
How many sheets of paper does the bundle have?

⟨Ans.⟩ _____ sheets

( 2 ) When measuring 40 sheets of paper, the weight is 48 g.
How many sheets of paper does the bundle have?

⟨Ans.⟩ _____ sheets

( 3 ) When the bundle is 1,000 sheets, what is the weight?

⟨Ans.⟩ _____ kg

**40**

# Proportions & Inverse Proportions Review 2

Level ★★★

Date        /        /

Name

Score
/100

■ The Answer Key is on page 95.

**1**  There is a rectangle with a length of 5 cm and a width of 8 cm.   10 points
If the area of this rectangle does not change and the width is
set to 4 cm, how many centimeters is the length?

⟨Ans.⟩                                cm

**2**  Christopher takes 25 minutes walking at 60 m per minute to go from   10 points
home to the library.
How long will it take, if he walks on the same road at a speed of
75 m/min?

⟨Ans.⟩                           minutes

**3**  Nancy is filling the bathtub with hot water.   10 points
It took 30 minutes to fill the bathtub at a rate of 6 ℓ per minute.
How long will it take if she fills the bathtub with hot water at a rate of
10 ℓ per minute?

⟨Ans.⟩                           minutes

**4**  It takes 4 hours to go to station A by the local train at 90 km/h.   10 points
However, the limited express train will arrive at station A in 1.5 hours.
How many kilometers is the speed per hour of the limited express train?

⟨Ans.⟩                              km/h

**5**  There is a parallelogram with a height of 12 cm and a base of 3 cm.   10 points
If not changing the area of this parallelogram and setting the height to 9 cm,
how many centimeters is the base?

⟨Ans.⟩                                cm

**6** There is a rectangle with a length of 4 cm and a width of 12 cm. 10 points

If the area of this rectangle does not change and the length is set to 6 cm, how many centimeters is the width?

〈Ans.〉 _____ cm

**7** Daniel takes 14 minutes to walk to the school at a speed of 65 m/min. 10 points

How long will it take him to arrive when he walks at a speed of 70 m/min?

〈Ans.〉 _____ minutes

**8** If putting water into the pool every 2 m³ per hour, it will be full in 15 hours. 10 points

How many cubic meters of water should be added per hour to fill it in 6 hours?

〈Ans.〉 _____ m³

**9** Matthew decided to go by bike to the neighboring town which takes 10 points

1.2 hours by bus at 45 km/h. If he tries to ride to the neighboring town in 3 hours, what is the speed per hour of his riding?

〈Ans.〉 _____ km/h

**10** Betty is cutting a drawing paper to make a flag. 10 points

She cuts it to 10 cm in length and 15 cm in width.

Dorothy is also cutting the drawing paper to make a flag of the same area.

When she cuts the width to 12.5 cm, how long should the length be?

〈Ans.〉 _____ cm

**41**

# Review 1

Date       /       /

Name

Level
★★★

Score
_____/100

■ The Answer Key is on page 96.

**1**  Debra has $5.
She bought 3 cakes of the same price and had $0.80 left.
What is the price of 1 cake?

10 points

⟨Ans.⟩  $ _____

**2**  James has $0.50 more than his brother John.
The sum of their money is $2.50.
How much money does John have?

10 points

⟨Ans.⟩  $ _____

**3**  Robert bought a combination of 15 pencils and color pencils.
Each pencil was $0.60 and each colored pencil was $0.45.
In total he paid, $8.70.
How many pencils and colored pencils did he buy, respectively?

10 points

⟨Ans.⟩ _____ pencils

⟨Ans.⟩ _____ colored pencils

**4**  Mary is 7 years old and her father is 41 years old.
How many years will it take for her father's age to be 3 times Mary's age.

10 points

⟨Ans.⟩ _____ years later

**5** Patricia left her home at a speed of 80 m/min walking toward the station.　15 points
After 10 minutes, her elder brother Robert departed from their home
by bicycle and chased Patricia at a speed of 240 m/min.
Find how many minutes after Robert departs from his house will he catch
up with Patricia.

⟨Ans.⟩ _____ minutes later

**6** There are 600 spectators in the movie theater, and 5% of them are children.　15 points
Several adults came out from the movie theater, so the number of children
is the theater becomes 12%.
Find how many adults came out from the movie theater.

⟨Ans.⟩ _____ adults

**7** There is 500 g of an 8% salt solution.　15 points
If this saline solution is diluted with water to create a 5% salt solution.
How many grams of water were added to the saline solution?

⟨Ans.⟩ _____ g

**8** In the triangle **ABC** in the figure on the right.　15 points
The size of ∠**B** is twice ∠**A**
and ∠**C** is 1.5 times ∠**A**.
Find the sizes of ∠**A**, ∠**B**,
and ∠**C**, respectively.

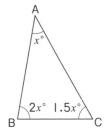

⟨Ans.⟩　∠A _____
⟨Ans.⟩　∠B _____
⟨Ans.⟩　∠C _____

# Review 2

**42**

Date / /

Name

■ The Answer Key is on page 96.

**1** The price for 5 apples is $1.60 higher than the price for 3 apples.
Find the price of 1 apple.

10 points

⟨Ans.⟩  $

**2** The price for 1 apple is $0.30 higher than the price for 1 orange.
If the price of 1 orange is added to the price of 1 apple it will be $2.10.
Find the price of 1 orange.

10 points

⟨Ans.⟩  $

**3** Jennifer bought 13 pieces of fruit in total.
The price of 1 pear was $1.40 and the price of 1 peach was $1.20.
She paid $16.60.
How many of each, pears and peaches did she buy?

10 points

⟨Ans.⟩                    pears

⟨Ans.⟩                    peaches

**4** Michael has $6 and his younger sister Linda has $1.20.
How much money does Michael have to give Linda, so that his money
is 3 times the amount Linda has?

10 points

⟨Ans.⟩  $

**5** David left his home at a speed of 80 m/min walking toward the station.
After 9 minutes, his younger sister, Elizabeth departed from their home
by bicycle and chased David at a speed of 200 m/min.
Find how many minutes after Elizabeth departs from her house she will
catch up with David.

15 points

⟨Ans.⟩ _____ minutes later

**6** There are 80 oranges and apples, and 5% of them are apples.
William takes out some of the oranges,
so apple make up 20% of the remaining pieces of fruit.
How many oranges did he take out?

15 points

⟨Ans.⟩ _____ oranges

**7** There is 500 g of an 8% salt solution.
In order to make the concentration of this saline solution 10%,
how many grams of water need to be evaporated?

15 points

⟨Ans.⟩ _____ g

**8** The sixth grade of the junior high school is divided into 3 classes,
and the total number of students in the sixth grade is 119.
The number of students in class **A** is 3 more people than the number of
students in class **B**, the number of students in class **B** is 2 less people than
the number of students in class **C**,
Find the number of students in class **A**, **B**, and **C**, respectively.

15 points

⟨Ans.⟩ class A _____ students

⟨Ans.⟩ class B _____ students

⟨Ans.⟩ class C _____ students

# 43 Review 3

Date / /

Name

Level
★★★

Score
/ 100

■ The Answer Key is on page 96.

**1** Richard's age will be 3 times his current age after 24 years. How old is he now?

10 points

⟨Ans.⟩ _____ years old

**2** Barbara has $ 1.50 more than her younger sister Susan. The sum of their money is $ 12.50. How much does Barbara have?

10 points

⟨Ans.⟩ $ _____

**3** The admission fee for the museum is $ 10 for 1 adult and $ 4 for 1 child. There were 90 visitors and the total admission fee was $ 798 on a certain day. Find the number of adults and children who entered that day.

10 points

⟨Ans.⟩ _____ adults

⟨Ans.⟩ _____ children

**4** The money Joseph has is twice the money Jessica has. After they both bought a book that costs $ 20, Joseph's money was 3 times as much as Jessica's. How much money did Joseph and Jessica each have at the beginning?

10 points

⟨Ans.⟩ Joseph $ _____

⟨Ans.⟩ Jessica $ _____

**5** Thomas has a savings of $15.60 now, plus a $0.80 savings every week. 15 points
Sarah has no savings, but decided to save $2 every week beginning
this week.
How many weeks will it take for Sarah's savings to be equal to the sum of
Thomas's savings?

⟨Ans.⟩ _____ weeks

**6** The total number of students at Susie's Junior High School is 1,000, 15 points
and 45% of them are girls.
As some boys moved, 50% of students became girls.
How many boys have moved to school?

⟨Ans.⟩ _____ boys

**7** There is 600 g of a 6% salt solution. 15 points
If this saline solution is diluted with water to make a 3% salt solution.
How many grams of water should be added?

⟨Ans.⟩ _____ g

**8** The money Charles, Margaret, and Karen have is a total of $185. 15 points
The money Charles has is $23 more than the money Margaret has.
The money Karen has is $12 less than the money Margaret has.
How much money do they each have respectively?

⟨Ans.⟩ Charles $ _____

⟨Ans.⟩ Margaret $ _____

⟨Ans.⟩ Karen $ _____

## (1) Review 1
pp 2,3

(1) **Ans.** red

(2) $\frac{4}{5} + 2\frac{3}{5} = 3\frac{2}{5}$  **Ans.** $3\frac{2}{5}$ ℓ

(3) $\frac{7}{9} \times 4 = \frac{28}{9} = 3\frac{1}{9}$  **Ans.** $3\frac{1}{9}$ ℓ $\left(\frac{28}{9} ℓ\right)$

(4) $1.12 \times 0.75 = 0.84$  **Ans.** 0.84 kg

(5) $35.5 \div 3.5 = 10.1\overset{.}{4}$  **Ans.** 10.1 km

(6) $(73.6 + 85.8 + 66.5) \div 3 = 75.3$  **Ans.** 75.3 pounds

(7) $18 \div 30 = 0.6$  **Ans.** 60%

(8) $280 \times 0.35 = 98$  **Ans.** 98 apples

(9) $2.4 \times 1.25 = 3$  **Ans.** $3

(10) $7.50 - 5.10 = 2.40, \quad 5 - 3 = 2$
$2.40 \div 2 = 1.20$
$5.10 - 1.20 \times 3 = 1.50$
**Ans.** Apple $1.50
**Ans.** Orange $1.20

## (2) Review 2
pp 4,5

(1) The least common multiple of 15 and 18 is 90.
$9:30 + 90$ minutes $= 11:00$
**Ans.** 11 AM

(2) $3\frac{1}{2} - 1\frac{1}{3} = 2\frac{1}{6}$  **Ans.** $2\frac{1}{6}$ gallons$\left(\frac{13}{6}$ gallons$\right)$

(3) $2\frac{1}{2} \div 6 = \frac{5}{2} \times \frac{1}{6} = \frac{5}{12}$  **Ans.** $\frac{5}{12}$ ℓ

(4) $2.5 \times 2.24 = 5.6$  **Ans.** 5.6 ℓ

(5) $14.4 \div 3.2 = 4.5$  **Ans.** 4.5 m

(6) $4.5$ kg $= 4500$ g, $\quad 4500 \div 75 = 60$
[Also, $75$ g $= 0.075$ kg, $\quad 4.5 \div 0.075 = 60$]
**Ans.** 60 oranges

(7) The West Elementary School : $540 \div 450 = 1.2$
The East Elementary School : $728 \div 560 = 1.3$
**Ans.** The East Elementary School

(8) $12 \div 32 = 0.375$  **Ans.** 37.5%

(9) $1.40 \div 0.35 = 4$  **Ans.** $4

(10) $20 \div (20 + 230) = 0.08$  **Ans.** 8%

## (3) Fractions 1
pp 6,7

(1) $\frac{3}{4} \times 5 = \frac{15}{4} = 3\frac{3}{4}$  **Ans.** $3\frac{3}{4}$ kg

(2) $\frac{7}{10} \div 3 = \frac{7}{30}$  **Ans.** $\frac{7}{30}$ kg

(3) $7 \times \frac{5}{6} = \frac{35}{6} = 5\frac{5}{6}$  **Ans.** $5\frac{5}{6}$ ℓ

(4) $\frac{1}{8} \times \frac{3}{7} = \frac{3}{56}$  **Ans.** $\frac{3}{56}$ kg

(5) $2 \times \frac{1}{4} = \frac{1}{2}$  **Ans.** $\frac{1}{2}$ pounds

(6) $10 \div \frac{3}{5} = \frac{50}{3} = 16\frac{2}{3}$  **Ans.** $16\frac{2}{3}$ ℓ

(7) $7 \div \frac{1}{6} = 42$  **Ans.** 42 pieces

(8) $3 \div \frac{3}{5} = 5$  **Ans.** 5 times

(9) $\frac{3}{4} \div \frac{1}{12} = 9$  **Ans.** 9 bags

(10) $\frac{3}{5} \div \frac{4}{5} = \frac{3}{4}$  **Ans.** $\frac{3}{4}$ times

## (4) Fractions 2
pp 8,9

(1) $\frac{3}{4} \div 5 = \frac{3}{20}$  **Ans.** $\frac{3}{20}$ kg

(2) $5 \times 1\frac{3}{4} = \frac{35}{4} = 8\frac{3}{4}$  **Ans.** $8.75

(3) $1\frac{7}{9} \times \frac{3}{4} = \frac{4}{3} = 1\frac{1}{3}$  **Ans.** $1\frac{1}{3}$ m²

(4) $3\frac{1}{5} \times \frac{5}{6} = \frac{8}{3} = 2\frac{2}{3}$  **Ans.** $2\frac{2}{3}$ kg $\left(\frac{8}{3}$ kg$\right)$

(5) $\frac{3}{8} \div \frac{2}{5} = \frac{15}{16}$  **Ans.** $\frac{15}{16}$ kg

(6) $15 \div 1\frac{2}{3} = 9$  **Ans.** 9 people

(7) $10 \div \frac{3}{2} = \frac{20}{3} = 6\frac{2}{3}$  **Ans.** $6\frac{2}{3}$ times $\left(\frac{20}{3}$ times$\right)$

(8) $1\frac{1}{4} \div \frac{5}{8} = 2$  **Ans.** 2 bags

(9) $1\frac{4}{5} \div \frac{2}{7} = \frac{63}{10} = 6\frac{3}{10}$  **Ans.** $6\frac{3}{10}$ times $\left(\frac{63}{10}$ times$\right)$

(10) $\frac{1}{5} \times 6 - \frac{1}{5} = 1$
$\left[$Also, $\frac{1}{5} \times (6-1) = 1\right]$  **Ans.** 1 kg

## (5) Fractions 3
pp 10,11

(1) $2\frac{2}{9} \div 6 = \frac{10}{27}$  **Ans.** $\frac{10}{27}$ feet

(2) $3 \times 3\frac{5}{6} = \frac{23}{2} = 11.5$  **Ans.** $11.50

(3) $1\frac{5}{7} \times 1\frac{1}{6} = 2$  **Ans.** 2 m

(4) $1\frac{3}{7} \times 2\frac{5}{8} = \frac{15}{4} = 3\frac{3}{4}$  **Ans.** $3\frac{3}{4}$ m $\left(\frac{15}{4}$ m$\right)$

(5) $3\frac{1}{8} \div 3\frac{3}{4} = \frac{5}{6}$  **Ans.** $\frac{5}{6}$ m²

(6) $18 \div 1\frac{1}{5} = 15$  **Ans.** 15 pieces

(7) $3 \div 2\frac{1}{4} = \frac{4}{3} = 1\frac{1}{3}$  **Ans.** $1\frac{1}{3}$ times $\left(\frac{4}{3}$ times$\right)$

⑧ $7\frac{1}{2} \div 1\frac{1}{4} = 6$      **Ans.** 6 pieces

⑨ $(1.80 + 2.40) \times \frac{5}{6} = 4.20 \times \frac{5}{6} = 3.50$

  $\left[\text{Also,}\ 1.80 \times \frac{5}{6} + 2.40 \times \frac{5}{6} = 3.50\right]$

                      **Ans.** $3.50

⑩ $\frac{5}{6} \times \left(0.8 + \frac{3}{5}\right) = \frac{5}{6} \times \left(\frac{4}{5} + \frac{3}{5}\right) = \frac{5}{6} \times \frac{7}{5} = \frac{7}{6} = 1\frac{1}{6}$

            **Ans.** $1\frac{1}{6}$ m² $\left(\frac{7}{6}\ \text{m}^2\right)$

---

## ⑥ Speed 1      pp 12, 13

1. $80 \div 2 = 40$      **Ans.** 40 km/h
2. $2700 \div 15 = 180$      **Ans.** 180 m/min
3. $120 \div 8 = 15$      **Ans.** 15 m/sec
4. $340 \times 60 = 20400$      **Ans.** 20400 m/min
5. $40 \times 3 = 120$      **Ans.** 120 km
6. $12 \div 3 = 4$      **Ans.** 4 hours
7. $54 \div \frac{3}{4} = 72$      **Ans.** 72 miles/h
8. $60 \times \frac{45}{60} = 45$      **Ans.** 45 km
9. $5 \times 60 = 300$      **Ans.** 300 km/h
10. $1500 \div 60 = 25$      **Ans.** 25 m/sec

---

## ⑦ Speed 2      pp 14, 15

1. 3 minutes = 180 seconds
   $72 \times 10 \div 180 = 4$      **Ans.** 4 m/sec
2. $900 \div 60 = 15$      **Ans.** 15 miles/min
3. $64 \times 2.5 = 160$      **Ans.** 160 km
4. 500 m/min = 0.5 km/min
   $0.5 \times 25 = 12.5$
   [Also, $500 \times 25 = 12500$, 12500 m = 12.5 km]
                   **Ans.** 12.5 km
5. $910 \div 65 = 14$      **Ans.** 14 minutes
6. $24 \div \frac{40}{60} = 36$      **Ans.** 36 km/h
7. $150 \div \frac{24}{60} = 375$      **Ans.** 375 m/min
8. 54 km = 54000 m
   $54000 \div 720 = 75$
   75 minutes = 1 hour 15 minutes
          **Ans.** 1 hour 15 minutes
9. $80 \times 1\frac{3}{4} = 140$      **Ans.** 140 miles
10. $\frac{15}{4} \times \frac{25}{60} = \frac{25}{16} = 1\frac{9}{16}$    **Ans.** $1\frac{9}{16}$ km $\left(\frac{25}{16}\ \text{km}\right)$

---

## ⑧ Speed 3      pp 16, 17

1. $30 \div 60 = 0.5$,   $0.5 \times 20 = 10$      **Ans.** 10 miles
2. 10:20 AM − 9:40 AM = 40 minutes
   $72 \div 60 = 1.2$,   $1.2 \times 40 = 48$      **Ans.** 48 miles
3. $12 \times 60 = 720$,   21.6 km = 21600 m
   $21600 \div 720 = 30$      **Ans.** 30 minutes
4. $6 \times 60 = 360$,   1.8 km = 1800 m
   $1800 \times 3 \div 360 = 15$      **Ans.** 15 minutes
5. $100 \div \frac{18}{60} = \frac{1000}{3} = 333\frac{1}{3}$

       **Ans.** $333\frac{1}{3}$ m/min $\left(\frac{1000}{3}\ \text{m/min}\right)$
6. 1 hour 20 minutes = 80 minutes = $\frac{80}{60}$ hour

   $1000 \div \frac{80}{60} = 750$      **Ans.** 750 km/h
7. 2 hour 15 minutes = 135 minutes = $\frac{135}{60}$ hour

   $100 \div \frac{135}{60} = \frac{400}{9} = 44\frac{4}{9}$

       **Ans.** $44\frac{4}{9}$ miles/h $\left(\frac{400}{9}\ \text{miles/h}\right)$
8. $880 \div 480 = \frac{11}{6} = 1\frac{5}{6}$

   $1\frac{5}{6}$ hour = 1 hour 50 minutes      **Ans.** 1 hour 50 minutes
9. $2 \div \frac{15}{4} = \frac{8}{15}$,   $\frac{8}{15}$ hour = 32 minutes

                **Ans.** 32 minutes
10. $11 \div \frac{55}{4} = \frac{4}{5}$,   $\frac{4}{5}$ hour = 48 minutes

                 **Ans.** 48 minutes

---

## ⑨ Algebraic Expressions 1      pp 18, 19

1. $x + 4 = 12$
   $x = 8$      **Ans.** 8 oranges
2. $x - 3 = 7$
   $x = 10$      **Ans.** 10 ounces
3. $x \times 3 = 1.80$
   $x = 0.60$      **Ans.** $0.60
4. $x \times 4 = 26$
   $x = 6.5$      **Ans.** 6.5 cm
5. $x \div 5 = 3.6$
   $x = 18$      **Ans.** 18 ounces
6. $x \times 9 = 45$
   $x = 5$      **Ans.** 5 cm
7. $x \times 8 = 72$
   $x = 9$      **Ans.** 9 cm
8. $6 \times x = 45$
   $x = 7.5$      **Ans.** 7.5 cm
9. $6 \times x = 33$
   $x = 5.5$      **Ans.** 5.5 cm

## ⑩ Algebraic Expressions 2 pp 20, 21

**①** $x \times 8 \times 4 = 160$
$\quad x = 5$       **Ans.** 5 cm

**②** $x \times 5 \times 4 = 800$
$\quad x = 40$       **Ans.** 40 g

**③** $x \times 6 \times 5 = 15$
$\quad x = 0.50$       **Ans.** \$0.50

**④** $x \times 4 \times 3 = 12$
$\quad x = 1$       **Ans.** \$1

**⑤** (1) $30 + x = y$
(2) $2 - x = y$
(3) $x - 2.5 = y$
(4) $2.60 \times x = y$
(5) $x \times 9 = y$
(6) $x \div 5 = y$

**⑥** (1) $x \times 3 = y$
(2) $4 \times 3 = 12, \ y = 12$       **Ans.** 12 cm
(3) $x \times 3 = 42$
$\quad x = 14$       **Ans.** 14 cm

## ⑪ Linear Equations 1 (Solving Equations) pp 22, 23

**①**
(1) $x = 6$
(2) $x = -4$
(3) $x = -6$
(4) $x = 0$
(5) $x = 4.8$
(6) $x = -0.5$
(7) $x = 5.5$
(8) $x = -4\frac{1}{3}$
(9) $x = 2\frac{1}{5}$
(10) $x = -6\frac{3}{4}$
(11) $x = \frac{5}{6}$
(12) $x = \frac{1}{12}$
(13) $x = 2\frac{1}{6}$
(14) $x = \frac{1}{4}$

**②**
(1) $x = 3$
(2) $x = -12$
(3) $x = -7$
(4) $x = 1\frac{1}{3}$
(5) $x = -8$
(6) $x = -48$
(7) $x = -2\frac{2}{3}$
(8) $x = 3\frac{1}{3}$
(9) $x = -9$
(10) $x = \frac{8}{15}$
(11) $x = -4\frac{4}{5}$
(12) $x = 2\frac{2}{3}$
(13) $x = \frac{5}{6}$
(14) $x = 1\frac{13}{42}$
(15) $x = -10\frac{1}{2}$
(16) $x = -\frac{18}{35}$

## ⑫ Linear Equations 2 (Solving Equations) pp 24, 25

**①**
(1) $x = 3$
(2) $x = 3$
(3) $x = -1\frac{5}{6}$
(4) $x = 3\frac{1}{5}$
(5) $x = -9$
(6) $x = 10$
(7) $x = -5$
(8) $x = -1$
(9) $x = \frac{1}{2}$
(10) $x = -\frac{1}{6}$
(11) $x = -30$
(12) $x = 2\frac{2}{5}$

**②**
(1) $x = 6$
(2) $x = 1\frac{3}{5}$
(3) $x = -5$
(4) $x = 1\frac{5}{13}$
(5) $x = 2\frac{1}{3}$
(6) $x = -18$
(7) $x = \frac{5}{6}$
(8) $x = 1\frac{13}{14}$
(9) $x = -2$
(10) $x = 2\frac{7}{9}$
(11) $x = 1\frac{1}{15}$
(12) $x = 3\frac{7}{17}$
(13) $x = \frac{20}{63}$
(14) $x = -\frac{82}{189}$
(15) $x = \frac{23}{45}$
(16) $x = -\frac{7}{10}$

## ⑬ Linear Equations 3 (Pattern 1) pp 26, 27

★ **TRY!**
(1) $0.80, \ x, \ 0.80x$
(2) $0.50$
(3) $4.50$
(4) $0.80x, \ 0.50, \ 4.50$       **Ans.** 5 donuts

**①** $x, \ 1.40, \ 6.20, \ 0.60x + 1.40, \ 6.20$       **Ans.** 8 pencils

**②** Let $x$ be the number of donuts.
$\quad 1.50x + 0.50 = 11$
$\qquad x = 7$       **Ans.** 7 donuts

**③** Let $x$ be the number of cakes.
$\quad 1.70x + 0.25 = 10.45$
$\qquad x = 6$       **Ans.** 6 cakes

## ⑭ Linear Equations 4 (Pattern 2) pp 28, 29

★ **TRY!**
(1) $8x, \ 1.20$
(2) $x, \ 1.80$
(3) $8x + 1.20, \ x + 1.80$       **Ans.** \$0.40

**①** $x, \ 0.70, \ x, \ 5.20, \ 6x + 0.70, \ x + 5.20$       **Ans.** \$0.90

**②** Let $x$ be the price of one pencil.
$\quad 3x + 3 = 4(x + 0.60)$
$\qquad x = 0.60$       **Ans.** \$0.60

**③** Let $x$ be the price of one pencil.
$\quad 5x + 3 = 2(3x + 2 \times 0.40)$
$\qquad x = 1.40$       **Ans.** \$1.40

## 15 Linear Equations 5 (Pattern 3) — pp 30,31

★ TRY!

(1) $x$, 0.40

(2) 5, $5x+0.40$, 0.60

**Ans.** $0.80

**1** $x$, 6, $50-(3x+6)$, 8

**Ans.** $12

**2** Let $x$ be the price of one colored pencil.
$10-(7x+2\times1.30)=1.10$
$x=0.90$ **Ans.** $0.90

**3** Let $x$ be the price of one colored pencil.
$10-(5x+3\times1.50)=2$
$x=0.70$ **Ans.** $0.70

## 16 Linear Equations 6 (Pattern 4) — pp 32,33

★ TRY!

(1) $10-x$

(2) $x$, $10-x$

(3) $x$, $10-x$, 11

(4) 5 stamps

(5) 5 stamps

**1** $12-x$, $1.50x$, $12-x$, $1.50x+1.60(12-x)$, 18.50

**Ans.** 7 peaches

**Ans.** 5 pears

**2** Let $x$ be the number of cream puffs.
The number of ice creams is $10-x$.
$1.50x+1.20(10-x)=13.80$
$x=6$
The number of ice creams is $10-6=4$.

**Ans.** 6 cream puffs

**Ans.** 4 ice creams

**3** Let $x$ be the number of bananas.
The number of apples is $20-x$.
$1.20x+1.40(20-x)=26$
$x=10$
The number of apples is $20-10=10$.

**Ans.** 10 bananas

**Ans.** 10 apples

## 17 Linear Equations 7 (Pattern 5) — pp 34,35

★ TRY!

(1) $x+14$

(2) $x$, $x+14$, 50

(3) 18, 14, 32

**Ans.** 18 sheets

**Ans.** Larry 32 sheets

**Ans.** John 18 sheets

**1** $x+60$, $x+(x+60)$, 180

**Ans.** Ruth 120 cm

**Ans.** Ann 60 cm

**2** Let $x$ be the number of girls.
The number of boy is $x+3$.
$x+(x+3)=41$
$x=19$
The number of boy is $19+3=22$.

**Ans.** Boys 22 students

**Ans.** Girls 19 students

**3** Let $x$ be the number of girls.
The number of boy is $x+4$.
$x+(x+4)=84$
$x=40$
The number of boy is $40+4=44$.

**Ans.** Boys 44 students

**Ans.** Girls 40 students

## 18 Linear Equations 8 (Pattern 6) — pp 36,37

★ TRY!

(1) 41, $x$, 13, $x$

(2) $41-x$, $13+x$, 10

**Ans.** 9 marbles

**1** $26-x$, $11+x$, $26-x$, $(11+x)+5$

**Ans.** 5 cards

**2** If Pamela passes $x$ cookies to Scott,
the Scott's cookies : $14+x$
the Pamela's cookies : $24-x$
$14+x=(24-x)+2$
$x=6$ **Ans.** 6 cookies

**3** If David passes $x$ cupcakes to Sydney,
the David's cupcakes : $28-x$
the Scott's cupcakes : $12+x$
$(28-x)+6=12+x$
$x=11$ **Ans.** 11 cupcakes

## 19 Linear Equations 9 (Pattern 7) — pp 38,39

★ TRY!

(1) $4x$, 10

(2) $6x$, 8

(3) $4x$, 10, $6x$, 8 **Ans.** 9 students

**1** $3x+15$, $5x-1$, $3x+15=5x-1$

**Ans.** 8 students

**2** Let $x$ be the number of children.
$3x-7=2x+14$
$x=21$ **Ans.** 21 children

**3** Let $x$ be the number of friends.
$4x+12=6x-4$
$x=8$ **Ans.** 8 friends

## 20 Linear Equations 10 (Pattern 8) — pp 40,41

★ TRY!

(1) $n$, 4, $n$, 12

(2) $n$, 4, $n$, 12

**Ans.** 4

**1** $3x+5$, $5x-9$, $3x+5=5x-9$

**Ans.** 7

**2** Let $x$ be the number on Brandon's jersey.
$2x+4=6x-8$
$x=3$ **Ans.** 3

**3** Let $x$ be the number on James' racecar.
$3x+7=6x-2$
$x=3$ **Ans.** 3

## 21 Linear Equations 11 (Pattern 9) pp 42,43

★ TRY!

(1) $x$, 15

(2) $x$, 15, 5, $x$, 5

(3) $x$, 5, 200       **Ans.** 13

① $8x$, 5, 3, 48, $8x=48$       **Ans.** 6

② $12x=7\times6+(x+2)$
$x=4$       **Ans.** 4

③ $7x=9\times5+(x-3)$
$x=7$       **Ans.** 7

## 22 Linear Equations 12 (Pattern 10) pp 44,45

★ TRY!

(1) $x$, 7      (3) $x$, $x$, 7    **Ans.** 6

(2) $x$      (4) 6, 10, 67    **Ans.** 67

① $10x+3$, $30+x$, $30+x$, $(30x+3)-27$       **Ans.** 63

② Let $x$ be the number of ones digit of the original natural number.
$10x+8=(80+x)-36$
$x=4$
The original number is $80+4=84$.       **Ans.** 84

③ Let $x$ be the number of ones digit of the original natural number.
$10x+5=(50+x)+18$
$x=7$
The original number is $50+7=57$.       **Ans.** 57

## 23 Linear Equations 13 (Pattern 11) pp 46,47

★ TRY!

(1) $x$, $x$      (2) $46+x$, $14+x$       **Ans.** 2 years later

① $12+x$, $44+x$, $44+x$, $12+x$       **Ans.** 4 years later

② The equation for the age of two people $x$ years later is,
$17+x=2(7+x)$
$x=3$       **Ans.** 3 years later

③ The equation for the age of two people $x$ years ago is,
$15-x=\dfrac{1}{4}(45-x)$
$x=5$       **Ans.** 5 years ago

## 24 Linear Equations 14 (Pattern 12) pp 48,49

★ TRY!

(1) 60, $x$, $70x$      (2) 60, $x$, $70x$       **Ans.** 12 minutes later

① $12+x$, $200x$, $12+x$, $200x$       **Ans.** 8 minutes later

② Let $x$ be the time until Christian's sister catches up to Christian.
$60(15+x)=210x$
$x=6$       **Ans.** 6 minutes later

③ Let $x$ be the time until Sophie's brother catches up to Sophie.
$40(6+x)=100x$
$x=4$       **Ans.** 4 minutes later

## 25 Linear Equations 15 (Pattern 13) pp 50,51

★ TRY!

(1) $70x$, $80x$      (2) $70x$, $80x$, 1800       **Ans.** 12 minutes later

① $240x$, $160x$, $240x+160x$, 2400       **Ans.** 6 minutes later

② Let $x$ be the time until the two meet first.
$10x+14x=12$
$x=\dfrac{1}{2}$       **Ans.** $\dfrac{1}{2}$ hours later
[Also, 0.5 hours later]

③ Let $x$ be the time until the two meet first.
$13x+7x=40$
$x=2$       **Ans.** 2 hours later

## 26 Linear Equations 16 (Pattern 14) pp 52,53

★ TRY!

(1) $200x$, $160x$      (2) $200x$, $160x$, 800       **Ans.** 20 minutes later

① $70x$, $170x$, $170x-70x$, 400       **Ans.** 4 minutes later

② Let $x$ be the speed per minute of Teresa's walk.
$300\times8-x\times8=2000$
$x=50$       **Ans.** 50 m/min

③ Let $x$ be the speed per minute of Thomas's walk.
$400\times15-x\times15=5100$
$x=60$       **Ans.** 60 m/min

## 27 Linear Equations 17 (Pattern 15) pp 54,55

★ TRY!

(1) $x$, $x$      (2) $\dfrac{x}{45}$, $\dfrac{x}{60}$       **Ans.** 3600 m

① $\dfrac{x}{4}$, $\dfrac{x}{5}$, $\dfrac{x}{4}-\dfrac{x}{5}$       **Ans.** 20 km

② Let $x$ be the distance from City A to City B.
$\dfrac{x}{6}-\dfrac{x}{14}=2$
$x=21$       **Ans.** 21 km

③ Let $x$ be the distance Arya´s house to her sister´s house.

$$\frac{x}{6}-\frac{x}{12}=2\frac{1}{3}$$
$$x=28$$

**Ans.** 28 km

## 28 Linear Equations 18 (Pattern 16)    pp·56,57

★ **TRY!**

(1) 140, $x$, 100

(2) $x$ : 100, 70 : 140

**Ans.** 50 ml

| How to Solve | $x\times140=100\times70$ |
|---|---|

$$x=100\times\frac{70}{140}=50$$

① 150, $x$, 400, $x$ : 400, 60 : 150

**Ans.** 160 g

② Let $x$ be the amount of grams of butter Julia needs.

$x$ : 350 = 80 : 200
$$x=140$$

**Ans.** 140 g

③ Let $x$ be the amount of ounces of milk Kyle needs.

$x$ : 700 = 260 : 400
$$x=455$$

**Ans.** 455 ounces

## 29 Linear Equations 19 (Pattern 17)    pp 58,59

★ **TRY!**

(1) 7

(2) 140, $x$

(3) $x$, 7:4

**Ans.** 80 sheets

| How to Solve | $140\times4=x\times7$ |
|---|---|

$$x=140\times\frac{4}{7}=80$$

① 9, 36, $x$, 36 : $x$, 9 : 4

**Ans.** 16 caramels

② Let $x$ be the number of marbles Judy´s sister receives.

84 : $x$ = 12 : 5
$$x=35$$

**Ans.** 35 marbles

③ Let $x$ be the number of crayons Niel´s brother receives.

56 : $x$ = 8 : 3
$$x=21$$

**Ans.** 21 crayons

## 30 Equations Review 1    pp 60,61

① (1) $2x+16=85-x$
$$x=23$$

**Ans.** 23

(2) $3x+5=2(x-2)+1$
$$x=-8$$

**Ans.** −8

② Let $x$ be the price of one notebook.

$12-10x=2(4-3x)$
$$x=1$$

**Ans.** $1

③ Let $x$ be the price of one pencil.
The price of one notebook is $(2.10-x)$.

$(2.10-x)+5x=5.10$
$$x=0.75$$

**Ans.** Notebook $1.35

**Ans.** Pencil $0.75

④ Let $x$ be the number of children.

$6x+23=9x+5$
$$x=6$$

**Ans.** 6 children

⑤ Let $x$ be the distance from point A to point B.

$$\frac{x}{6}+\frac{x}{4}=7.5$$
$$x=18$$

**Ans.** 18 km

⑥ Let $x$ be the number of the population of City A.

$1.08x=16200$
$$x=15000$$

**Ans.** 15000 people

⑦ 1 kg = 1000 g

salt: $1000\times\frac{3}{100}=30$

water: $1000-30=970$

**Ans.** Salt 30 g

**Ans.** Water 970 g

⑧ Let $x$ be the amount of money Randy had initially.

$$x-\frac{1}{4}x-\left(1-\frac{1}{4}\right)x\times\frac{2}{5}=5.40$$
$$x=12$$

**Ans.** $12

## 31 Equations Review 2    pp 62,63

① Let $x$ be the price of one notebook.

$2-2x=0.20$
$$x=0.90$$

**Ans.** $0.90

② Let $x$ be the amount of money Abigail and Natalie received from their mother.

$6+x=2(1+x)$
$$x=4$$

**Ans.** $4

③ Let $x$ be the price of one orange.
The price of one apple is $(2-x)$.

$5x+7(2-x)=12.40$
$$x=0.80$$

**Ans.** Orange $0.80

**Ans.** Apple $1.20

④ Let $x$ be the number of students.

$5x+45=8x+6$
$$x=13$$

**Ans.** 13 students

**Ans.** 110 colored pencils

⑤ Catches up to Jane. Let $x$ be the time until Gregory.

$200x=80x+600$
$120x=600$
$$x=5$$

**Ans.** 5 minutes later

⑥ Let $x$ be the original cost of the item.

$1.20\times x=7.20$
$$x=6$$

**Ans.** $6

⑦ $500\times\frac{10}{100}=50$

Let $x$ be the amount of water to add.

$(500+x)\times\frac{4}{100}=50$
$(500+x)\times4=5000$
$$x=750$$

**Ans.** Salt 50 g

**Ans.** Water 750 g

**8**  Let $x$ be the amount of money Virginia had initially.
$$x - \frac{1}{4}x - 5 = \frac{1}{2}x + 0.40$$
$$x = 21.60 \qquad \textbf{Ans.} \ \ \$21.60$$

**(32) Proportions & Inverse Proportions 1**  pp 64,65

(1) ① $= y$, ② $=$ a function
(2) ① $= 0.50x$, ② $=$ proportional, ③ $= 0.50$
(3) ① $= 1$, ② $= 2$
(4) ① $=$ a straight line, ② $=$ the origin
(5) ① $= 36$, ② $=$ inversely proportional, ③ $= 36$
(6) ① $=$ hyperbolic curve, ② $= 3$, ③ $= -2$

**(33) Proportions & Inverse Proportions 2 (Pattern 1)**  pp 66,67

★ **TRY!**
(1) $kx$
(2) $1$, $100$, $\dfrac{1}{100}$
(3) $13$  **Ans.** $13$ cm

**1**  $y = kx$, $300$, $540$, $540$, $300$, $\dfrac{9}{5}$, $\dfrac{9}{5}$
**Ans.** $81$ g

**How to Solve**  $\dfrac{9}{5} \times 45 = 81$

**2**  If the weight of $x$ m of the wire is $y$ g, since $y$ is proportional to $x$, the expression by using $k$ is represented as $y = kx$.
When $x = 5$, $y = 90$.
From $k = \dfrac{90}{5} = 18$, $y = 18x$.
The weight of $3$ m of the wire is $18 \times 3 = 54$ (g).
**Ans.** $54$ g

**3**  If the weight of $x$ ft of the rope is $y$ pounds, since $y$ is proportional to $x$, the expression by using $k$ is represented as $y = kx$.
When $x = 8$, $y = 120$.
From $k = \dfrac{120}{8} = 15$, $y = 15x$.
The weight of $5$ ft of the rope is $15 \times 5 = 75$ (pounds).
**Ans.** $75$ pounds

**(34) Proportions & Inverse Proportions 3 (Pattern 2)**  pp 68,69

★ **TRY!**
(1) $kx$
(2) $225$, $\dfrac{45}{8}$, $\dfrac{45}{8}$
(3) $315$  **Ans.** $315$ cm²

**1**  $y = kx$, $24$, $224$, $224$, $24$, $\dfrac{28}{3}$, $\dfrac{28}{3}$
**Ans.** $420$ cm²

**How to Solve**  $\dfrac{28}{3} \times 45 = 420$

**2**  If the area of $x$ g of the cardboard is $y$ cm², since $y$ is proportional to $x$, the expression by using $k$ is represented as $y = kx$.
When $x = 48$, $y = 32 \times 32 = 1024$.
From $k = \dfrac{1024}{48} = \dfrac{64}{3}$, $y = \dfrac{64}{3}x$.
The area of $75$ g of the cardboard is $\dfrac{64}{3} \times 75 = 1600$ (cm²).
**Ans.** $1600$ cm²

**3**  If the area of $x$ g of the cardboard is $y$ cm², since $y$ is proportional to $x$, the expression by using $k$ is represented as $y = kx$.
When $x = 36$, $y = 25 \times 20 = 500$.
From $k = \dfrac{500}{36} = \dfrac{125}{9}$, $y = \dfrac{125}{9}x$.
The area of $81$ g of the cardboard is $\dfrac{125}{9} \times 81 = 1125$ (cm²).
**Ans.** $1125$ cm²

**(35) Proportions & Inverse Proportions 4 (Pattern 3)**  pp 70,71

★ **TRY!**
(1) Distance
(2) $80$
(3) Speed
(4) $80$, $80$
**Ans.** $y = 80x$

**1**  $1200$, $24$, $50$, $50$, $50$  **Ans.** $y = 50x$

**2**  The speed when walking $15$ minutes for the distance of $1200$ m is $1200 \div 15 = 80$ (m/min).
Since $y$ is proportional to $x$, the distance of $y$ m when walking for $x$ minutes with the speed of $80$ m/min. is $y = 80x$.
**Ans.** $y = 80x$

**3**  Suppose that Dylan leaves Town A and $x$ hours later, it is $y$ km away from Town A.
The speed when walking $3$ hours for the distance of $12$ km is $12 \div 3 = 4$ (km /h).
Since $y$ is proportional to $x$, the distance of $y$ km when walking for $x$ hours with the speed of $4$ km/h is $y = 4x$.
So, $2$ hours after Dylan leaves Town A, he is at $4 \times 2 = 8$ (km).
**Ans.** $8$ km

**(36) Proportions & Inverse Proportions 5 (Pattern 4)**  pp 72,73

★ **TRY!**
(1) $x$, $y$
(2) $12$, $8$
(3) $96$, $96$, $x$  **Ans.** $y = \dfrac{96}{x}$

**1**  $y$, $18$, $10$  **Ans.** $y = \dfrac{180}{x}$

**(2)** $x \times y = 16 \times 20$

If $y$ is represented by the expression of $x$, $y = \dfrac{320}{x}$.

When the weight on the right side is $8$ g, the distance from the fulcrum is obtained by substituting $x = 8$ in the above equation.

$y = \dfrac{320}{8} = 40$

**Ans.** 40 cm

---

### (37) Proportions & Inverse Proportions 6 (Pattern 5)  pp 74,75

★ **TRY!**

(1) ×

(2) 180, $x$, $y$

(3) 180, $x$ **Ans.** $y = \dfrac{180}{x}$

**(1)** ×, $y$, $\dfrac{120}{x}$, 120, 8 **Ans.** 8 tulip bulbs

**(2)** If the number of rows is $x$ rows and the number of bulbs planted in 1 row is $y$,

$x \times y = 270$

If $y$ is represented by the expression of $x$, $y = \dfrac{270}{x}$

If $x$ is 18, the value of $y$ is, $y = \dfrac{270}{18} = 15$

**Ans.** 15 tulip bulbs

**(3)** If the number of rows is $x$ rows and the number of bulbs planted in 1 row is $y$,

$x \times y = 560$

If $y$ is represented by the expression of $x$, $y = \dfrac{560}{x}$

If $x$ is 14, the value of $y$ is, $y = \dfrac{560}{14} = 40$

**Ans.** 40 rose bushes

---

### (38) Proportions & Inverse Proportions 7 (Pattern 6)  pp 76,77

★ **TRY!**

(1) Time

(2) 48

(3) 48, $x$ **Ans.** $y = \dfrac{48}{x}$

**(1)** Time, $y$, $\dfrac{63}{x}$, 63, 21 **Ans.** 21 minutes

**(2)** If water is added at the rate of $x$ ℓ per minute and it becomes full in $y$ minutes,

$x \times y = 6 \times 10$

If $y$ is represented by the expression of $x$, $y = \dfrac{60}{x}$

If $x$ is 5, the value of $y$ is, $y = \dfrac{60}{5} = 12$

**Ans.** 12 minutes

---

**(3)** If water is added at the rate of $x$ ℓ per minute and it becomes full in $y$ minutes,

$x \times y = 36 \times 68$

If $y$ is represented by the expression of $x$, $y = \dfrac{2448}{x}$

If $x$ is 24, the value of $y$ is, $y = \dfrac{2448}{24} = 102$

**Ans.** 102 minutes

---

### (39) Proportions & Inverse Proportions Review 1  pp 78,79

**(1)** (1) $12 \div 5 = \dfrac{12}{5}$

$60 \times \dfrac{12}{5} = 144$ **Ans.** 144 km

(2) $60 \div 5 = 12$  $12 \times 12 = 144$ **Ans.** 144 km

**(2)** Let $x$ be the weight of the same bar of iron if it $24$ cm³.

$8 : 72 = 24 : x$

$x = 216$ **Ans.** 216 g

**(3)** Let $x$ be the volume of water will run out of the tap in 12 minutes.

$5 : 45 = 12 : x$

$x = 108$ **Ans.** 108 ℓ

**(4)** Let $x$ be the weight of 24 m of the wire.

$6 : 85 = 24 : x$

$x = 340$ **Ans.** 340 g

**(5)** (1) $50 \div 20 = 0.4$ **Ans.** 0.4 g

(2) Let $x$ be the weight of 70 nails.

$50 : 20 = 70 : x$

$x = 28$ **Ans.** 28 g

(3) $0.4 \,(g) \times 110 \,(nails) = 44 \,(g)$ **Ans.** 44 g

**(6)** (1) Let $x$ be the number of the papers.

$1 : 88 = 5 : x$

$x = 440$ **Ans.** 440 sheets

(2) Let $x$ be the number of the papers.

$40 : 48 = x : 528$

$x = 440$ **Ans.** 440 sheets

(3) Let $x$ be the weight of 1000 sheets.

$440 : 528 = 1000 : x$

$x = 1200$

$1200$ g $= 1.2$ kg **Ans.** 1.2 kg

---

### (40) Proportions & Inverse Proportions Review 2  pp 80,81

**(1)** If the length of the rectangle is $x$ cm and the width of the rectangle is $y$ cm,

$x \times y = 5 \times 8 = 40$

If $y$ is represented by the expression of $x$, $y = \dfrac{40}{x}$

If $x$ is 4, the value of $y$ is, $y = \dfrac{40}{4} = 10$

**Ans.** 10 cm

**2** If the Christopher's walking speed is $x$ m/min and the time for arriving to the library is $y$ minutes,

$x \times y = 60 \times 25 = 1500$

If $y$ is represented by the expression of $x$, $y = \dfrac{1500}{x}$

If $x$ is 75, the value of $y$ is, $y = \dfrac{1500}{75} = 20$

**Ans.** 20 minutes

**3** If the amount of hot water to come out in 1 minute is $x$ ℓ and the time that the bathtub is filled with hot water is $y$ minutes,

$x \times y = 6 \times 30 = 180$

If $y$ is represented by the expression of $x$, $y = \dfrac{180}{x}$

If $x$ is 10, the value of $y$ is, $y = \dfrac{180}{10} = 18$

**Ans.** 18 minutes

**4** If the speed of the train is $x$ km/h and the time for arriving at station A is $y$ hours,

$x \times y = 90 \times 4 = 360$

If $y$ is represented by the expression of $x$, $y = \dfrac{360}{x}$

If $x$ is 1.5, the value of $y$ is, $y = \dfrac{360}{1.5} = 240$

**Ans.** 240 km/h

**5** If the height of the parallelogram is $x$ cm and the base is $y$ cm,

$x \times y = 12 \times 3 = 36$

If $y$ is represented by the expression of $x$, $y = \dfrac{36}{x}$

If $x$ is 9, the value of $y$ is, $y = \dfrac{36}{9} = 4$

**Ans.** 4 cm

**6** If the length of the rectangle is $x$ cm and the width of the rectangle is $y$ cm,

$x \times y = 4 \times 12 = 48$

If $y$ is represented by the expression of $x$, $y = \dfrac{48}{x}$

If $x$ is 6, the value of $y$ is, $y = \dfrac{48}{6} = 8$

**Ans.** 8 cm

**7** If the Daniel's walking speed is $x$ m/min and the time for arriving to the school is $y$ minites,

$x \times y = 14 \times 65 = 910$

If $y$ is represented by the expression of $x$, $y = \dfrac{910}{x}$

If $x$ is 70, the value of $y$ is, $y = \dfrac{910}{70} = 13$

**Ans.** 13 minutes

**8** If the time that the pool is filled with water is $x$ hours and the amount of water to come out in 1 hour is $y$ m³,

$x \times y = 15 \times 2 = 30$

If $y$ is represented by the expression of $x$, $y = \dfrac{30}{x}$

If $x$ is 6, the value of $y$ is, $y = \dfrac{30}{6} = 5$

**Ans.** 5 m³

**9** If the time for arriving at the neighboring town is $x$ hours and the speed of the bike is $y$ km/h,

$x \times y = 1.2 \times 45 = 54$

If $y$ is represented by the expression of $x$, $y = \dfrac{54}{x}$

If $x$ is 3, the value of $y$ is, $y = \dfrac{54}{3} = 18$

**Ans.** 18 km/h

**10** If the width of the drawing paper is $x$ cm and the length of the paper is $y$ cm,

$x \times y = 15 \times 10 = 150$

If $y$ is represented by the expression of $x$, $y = \dfrac{150}{x}$

If $x$ is 12.5, the value of $y$ is, $y = \dfrac{150}{12.5} = 12$

**Ans.** 12 cm

## (41) Review 1 — pp 82, 83

1. $1.40
2. $1
3. 13 pencils, 2 colored pencils
4. 10 years later
5. 5 minutes later
6. 350 adults
7. 300 g
8. ∠A 40°, ∠B 80°, ∠C 60°

## (42) Review 2 — pp 84, 85

1. $0.80
2. $0.90
3. 5 pears, 8 peaches
4. $0.60
5. 6 minutes later
6. 60 oranges
7. 100 g
8. Class A 41 students
   Class B 38 students
   Class C 40 students

## (43) Review 3 — pp 86, 87

1. 12 years old
2. $7
3. 73 adults, 17 children
4. Joseph $80, Jessica $40
5. 13 weeks
6. 100 boys
7. 600 g
8. Charles $81
   Margaret $58
   Karen $46

## KUMON MATH WORKBOOKS

**Grades 6-8**

# Word Problems
## Workbook II

### Table of Contents

KUM☺N

# Simultaneous Linear Equations 1
## (Solving Equations)

Level

Date        /        /

Name

Score

/100

■ The Answer Key is on page 184.

**1** **Solve each equation.**

6 points per question

(1) $\begin{cases} 3x + 5y = 29 & \cdots ① \\ 9x - 2y = 19 & \cdots ② \end{cases}$

**Hint**   ① × 3

⟨Ans.⟩   $x =$          $y =$

(2) $\begin{cases} 5x + 7y = 3 \\ 3x + 14y = 6 \end{cases}$

⟨Ans.⟩   $x =$          $y =$

(3) $\begin{cases} 5x + 4y = 87 \\ 3x + y = 41 \end{cases}$

⟨Ans.⟩   $x =$          $y =$

(4) $\begin{cases} 8x - 3y = 16 \\ 2x - y = 4 \end{cases}$

⟨Ans.⟩   $x =$          $y =$

(5) $\begin{cases} 3x - 2y = 0.6 & \cdots ① \\ 4x + 3y = 2.5 & \cdots ② \end{cases}$

**Hint**   ① × 3   $9x - 6y = 1.8$
           ② × 2   $8x + 6y = 5$

⟨Ans.⟩   $x =$          $y =$

(6) $\begin{cases} -5x + 4y = -8.7 \\ 3x - y = 4.1 \end{cases}$

⟨Ans.⟩   $x =$          $y =$

(7) $\begin{cases} 3x + 2y = 0.6 \\ 4x - 3y = 2.5 \end{cases}$

⟨Ans.⟩   $x =$          $y =$

(8) $\begin{cases} 2x + 3y = 1.2 \\ 3x - 4y = 0.1 \end{cases}$

⟨Ans.⟩   $x =$          $y =$

## 2 Solve each equation.

( 1 )–( 4 ) 6 points per question  ( 5 )–( 8 ) 7 points per question

(1) $\begin{cases} 2x + 5y = x - 2y - 6 & \cdots① \\ 8x + y = 5x - y + 1 & \cdots② \end{cases}$

**Hint**   From ①, $x + 7y = -6$
From ②, $3x + 2y = 1$

⟨**Ans.**⟩   $x =$              $y =$

(2) $\begin{cases} 3x - y + 1 = 2x + 4 \\ 4x - 3y - 6 = 2y + 5 \end{cases}$

⟨**Ans.**⟩   $x =$              $y =$

(3) $\begin{cases} x = 2y \\ x - 4 = 3(y - 2) \end{cases}$

⟨**Ans.**⟩   $x =$              $y =$

(4) $\begin{cases} y = 3x \\ y - 3 = 5(x - 1) \end{cases}$

⟨**Ans.**⟩   $x =$              $y =$

(5) $\begin{cases} \dfrac{x}{3} + \dfrac{y}{2} = \dfrac{4}{3} & \cdots① \\ 2x = y & \cdots② \end{cases}$

**Hint**   From ①× 6, $2x + 3y = 8$
From ②, $2x - y = 0$

⟨**Ans.**⟩   $x =$              $y =$

(6) $\begin{cases} \dfrac{x}{3} - \dfrac{y}{2} = \dfrac{4}{3} \\ 2x = -y \end{cases}$

⟨**Ans.**⟩   $x =$              $y =$

(7) $\begin{cases} x + 1 = \dfrac{y}{4} \\ 5x = y + 1 \end{cases}$

⟨**Ans.**⟩   $x =$              $y =$

(8) $\begin{cases} x + 1 = \dfrac{y}{2} \\ 5x = 2y + 1 \end{cases}$

⟨**Ans.**⟩   $x =$              $y =$

# Simultaneous Linear Equations 2 (Solving Equations)

**2**

Level

Date / /

Name

Score

/100

■ The Answer Key is on page 184.

**1** **Solve each equation.**

6 points per question

(1) $\begin{cases} y = 3x - 1 & \cdots ① \\ 5x - y = 7 & \cdots ② \end{cases}$

**Hint** Substitute ② for ①,
$5x - (3x - 1) = 7$

⟨**Ans.**⟩ $x =$ $y =$

(2) $\begin{cases} y = 2x - 1 \\ x + y = 8 \end{cases}$

⟨**Ans.**⟩ $x =$ $y =$

(3) $\begin{cases} y = 2x + 1 \\ 3x + y = 16 \end{cases}$

⟨**Ans.**⟩ $x =$ $y =$

(4) $\begin{cases} y = -2x - 3 \\ 2x - y = 7 \end{cases}$

⟨**Ans.**⟩ $x =$ $y =$

(5) $\begin{cases} 2y = 3x - 5 & \cdots ① \\ 5x - 2y = 11 & \cdots ② \end{cases}$

**Hint** Substitute ② for ①,
$5x - (3x - 5) = 11$

⟨**Ans.**⟩ $x =$ $y =$

(6) $\begin{cases} 4x + 9y = 24 \\ 4x = 8 - 5y \end{cases}$

⟨**Ans.**⟩ $x =$ $y =$

(7) $\begin{cases} \dfrac{1}{2}x = 2 - y \\ \dfrac{1}{2}x - 3y = -2 \end{cases}$

⟨**Ans.**⟩ $x =$ $y =$

(8) $\begin{cases} \dfrac{1}{3}y = 2x - 7 \\ 3x + \dfrac{1}{3}y = 13 \end{cases}$

⟨**Ans.**⟩ $x =$ $y =$

## 2 Solve each equation.

( 1 )–( 4 ) 6 points per question  ( 5 )–( 8 ) 7 points per question

(1) $\begin{cases} 2x = y - 1 & \cdots ① \\ 3x - y = 1 & \cdots ② \end{cases}$

**Hint**  From ①, $y = 2x + 1$ $\cdots ③$
Substitute ② for ③,
$3x - (2x + 1) = 1$

$\langle$**Ans.**$\rangle$  $x =$ _____  $y =$ _____

(5) $\begin{cases} 2x + 3y = 7 & \cdots ① \\ 3x + 4y = 10 & \cdots ② \end{cases}$

**Hint**  From ①, $2x = -3y + 7$
$$x = \frac{-3y + 7}{2} \quad \cdots ③$$
Substitute ② for ③,
$$3\left(\frac{-3y + 7}{2}\right) + 4y = 10$$

$\langle$**Ans.**$\rangle$  $x =$ _____  $y =$ _____

(2) $\begin{cases} 4x + y = 10 \\ 3x - 2y = 2 \end{cases}$

$\langle$**Ans.**$\rangle$  $x =$ _____  $y =$ _____

(6) $\begin{cases} 3x + 4y = 4 \\ 2x - 3y = 14 \end{cases}$

$\langle$**Ans.**$\rangle$  $x =$ _____  $y =$ _____

(3) $\begin{cases} 2x - 3y = 8 \\ 2x = y \end{cases}$

$\langle$**Ans.**$\rangle$  $x =$ _____  $y =$ _____

(7) $\begin{cases} 5x + 2y = 18 \\ 13x + 3y = 49 \end{cases}$

$\langle$**Ans.**$\rangle$  $x =$ _____  $y =$ _____

(4) $\begin{cases} 2y = 3x - 2 \\ 4 = 2y - x \end{cases}$

$\langle$**Ans.**$\rangle$  $x =$ _____  $y =$ _____

(8) $\begin{cases} 5y = 10x - 15 \\ 3y - 2x = 3 \end{cases}$

$\langle$**Ans.**$\rangle$  $x =$ _____  $y =$ _____

**3**

Level ☆☆

Date / /

Name

Score

/100

■ The Answer Key is on page 184.

## Pattern 1 — Fee

● The admission fee for the museum is $15 for 2 adults and 1 junior high school student, and $12 for 1 adult and 2 junior high school students.
What are the admission fees for 1 adult and 1 junior high school student, respectively?

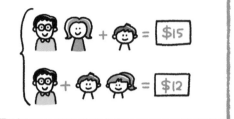

★ **TRY!** — **Fill in the blanks provided and solve for the answer.**      *15 points per question*

( 1 )  Assume that, the admission fee for 1 adult is $x$,

and the fee for 1 [                    ] is $y$.

( 2 )  If the admission fee for 2 adults and 1 junior high school student is represented by $x$ and $y$,

[                    ] ($)

If the admission fee for 1 adult and 2 junior high school students is represented by $x$ and $y$,

[                    ] ($)

( 3 )  The simultaneous equations for the admission fees are as follows.

The admission fee for 2 adults and 1 junior high school student

[                    ] = 15 ($)

The admission fee for 1 adult and 2 junior high school students

[                    ] = 12 ($)

Solve the above simultaneous equations.

$x = $ [          ] ,   $y = $ [          ]

⟨**Ans.**⟩  1 adult     $ _____

⟨**Ans.**⟩  1 student   $ _____

## Hint

Because the answer calls for two quantities, two equations are necessary.
The two equations can be solved simultaneously.
After solving the equations, make sure that the values found for $x$ and $y$ answer the question.

## Key Point

It is fundamental to represent two quantities as $x$ and $y$ in an equation.

And create two equations.

**1** The admission fee for a museum is $26 for 2 adults and 3 children, and $15 for 1 adult and 2 children.
What are the admission fees for 1 adult and 1 child, respectively?
Fill in the blanks provided and solve for the answer.

*15 points*

Assume that the admission fee for 1 adult is $$x$,

and the fee for 1 ⬚ is $$y$.

Represent the admission fee for 2 adults
and 3 children as $x$ and $y$,

⬚ ($)

Represent the admission fee for 1 adult
and 2 children as $x$ and $y$,

⬚ ($)

The simultaneous equations for the admission fees are as follows,

⬚

Solve the above simultaneous equations.

⟨**Ans.**⟩ 1 adult $ _____

⟨**Ans.**⟩ 1 child $ _____

**2** The admission fee for an amusement park is $26 for 2 adults and 1 children, and $28 for 1 adult and 3 children.
What are the admission fees for 1 adult and 1 child, respectively?

*20 points*

⟨**Ans.**⟩ 1 adult $ _____

⟨**Ans.**⟩ 1 child $ _____

**3** The admission fee for a music concert is $68 for 2 adults and 2 children, and the fee is $82 for 1 adult and 5 children.
What are the admission fees for 1 adult and 1 child, respectively?

*20 points*

⟨**Ans.**⟩ 1 adult $ _____

⟨**Ans.**⟩ 1 child $ _____

# Simultaneous Linear Equations 4 (Pattern 2)

Date / /

Name

Score

/100

■ The Answer Key is on page 184.

## Pattern 2 — Price

● The total price of 5 loaves of bread and 4 donuts is $6.40, and the total price of 6 loaves of bread and 3 donuts is $6.60.
What are the prices of 1 loaf of bread and 1 donut, respectively?

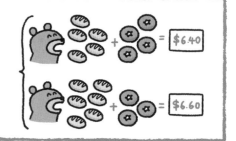

★ **TRY!** — Fill in the blanks provided and solve for the answer.

15 points per question

(1) Assume the price of 1 loaf of bread is $x$,

and the price of 1 [＿＿＿＿] is $y$.

(2) If the total price of 5 loaves of bread and 4 donuts is represented by $x$ and $y$,

[＿＿＿＿] ($)

If the total price of 6 loaves of bread and 3 donuts is represented by $x$ and $y$,

[＿＿＿＿] ($)

(3) The simultaneous equations for the total price of loaves of bread and donuts,

$$\begin{cases} \text{Total price for 5 loaves of bread and 4 donuts} \\ \boxed{\phantom{xxxx}} = 6.4 \ (\$) \\ \text{Total price for 6 loaves of bread and 3 donuts} \\ \boxed{\phantom{xxxx}} = 6.6 \ (\$) \end{cases}$$

Solve the above simultaneous equations.

$x =$ [＿＿＿], $y =$ [＿＿＿]

⟨**Ans.**⟩ 1 bread   $ ＿＿＿＿＿＿

⟨**Ans.**⟩ 1 donut   $ ＿＿＿＿＿＿

### Hint

$$\begin{cases} 15x + 12y = 19.2 & \cdots(1) \\ 24x + 12y = 26.4 & \cdots(2) \end{cases}$$

By multiplying both sides of the upper equation by 3 and both sides of the lower equation by 4, they can be solved as follows.

$(1) - (2) : -9x = -7.2$

**1** The total price of 3 roses and 4 lilies is $13.20,      15 points
and the total price of 2 roses and 5 lilies is $13.
What are the prices of 1 rose and 1 lily, respectively?
Fill in the blanks provided and solve for the answer.

Assume the price of 1 rose is $$x$$,

and the price for 1 [     ] is $$y$$.

Represent the price of 3 roses and 4 lilies by $x$ and $y$,     [     ] ($)

Represent the price of 2 roses and 5 lilies by $x$ and $y$,     [     ] ($)

The simultaneous equations for the total price of roses and lilies are as follows,

[     ]

Solve the above simultaneous equations.     ⟨**Ans.**⟩   1 rose    $

⟨**Ans.**⟩   1 lily    $

**2** The total price of 5 tulips and 4 cosmoses is $20.50,      20 points
and the total price of 4 tulips and 3 cosmoses is $16.
How much are the prices of 1 tulip and 1 cosmos, respectively?

⟨**Ans.**⟩   1 tulip    $

⟨**Ans.**⟩   1 cosmos    $

**3** The total cost of 7 sunflowers and 8 daffodils is $41.60,      20 points
and the total cost of 5 sunflowers and 4 daffodils is $25.60.
What are the costs of 1 sunflower and 1 daffodil, respectively?

⟨**Ans.**⟩   1 sunflower   $

⟨**Ans.**⟩   1 daffodil    $

# Simultaneous Linear Equations 5 (Pattern 3)

Level ☆☆

■ The Answer Key is on page 184.

## Pattern 3 — Number of Items

● Dylan bought 11 desserts, some of which were cream puffs that cost $1.30 each and some of which were puddings which cost $1 each.
He paid $12.80. How many of each cream puffs or puddings did he buy?

★ **TRY!** — Fill in the blanks provided and solve for the answer.

15 points per question

(1) Assume that Dylan bought $x$ number of cream puffs

and $y$ number of [ ].

(2) If the sum of cream puffs and puddings is represented by $x$ and $y$,

[ ] (pieces)

If the total price of $x$ pieces of cream puffs and $y$ pieces of puddings is represented by $x$ and $y$,

[ ] ($)

(3) The simultaneous equations for the sum of items and total price is,

$$\begin{cases} \boxed{\text{The sum of items} \quad\quad} = 11 \text{ (pieces)} \\ \boxed{\text{Total price} \quad\quad} = 12.8 \text{ ($)} \end{cases}$$

Solve the above simultaneous equations.

$x = $ [ ] , $y = $ [ ]

⟨Ans.⟩ _____ cream puffs

⟨Ans.⟩ _____ puddings

## Hint

This question can be solved with just one unknown value.
If there are $x$ cream puffs, the puddings are $11 - x$ (pieces),
so the equation for the price can be obtained by setting up
such an expression as $\$1.3x + \$1 \times (11 - x) = \$12.80$.

**Key Point**

"Sum of items"
"Total price"    Simultaneous equations can be created with these two equations.

**1** Theresa bought 9 pieces of fruit, some apples for $1.50 each and some peaches for $1.80 each. She paid $15 in total.
**How many apples and peaches did she buy?**
**Fill in the blanks provided and solve for the answer.**

If Theresa bought $x$ pieces of apples

and $y$ pieces of [       ].

If the sum of apples and peaches
is represented by $x$ and $y$,     [       ] (pieces)

If the total price of apples and peaches
is represented by $x$ and $y$,     [       ] ($)

The simultaneous equations for the sum of pieces
and the total price of apples and peaches is as follows,

[       ]

Solve the above simultaneous equations.   ⟨Ans.⟩ _____ apples
⟨Ans.⟩ _____ peaches

**2** Ralph bought 5 pears for $2.40 each and some persimmons for $1.50 each. He paid $9.30.
**How many pears and persimmons did he buy?**

20 points

⟨Ans.⟩ _____ pears
⟨Ans.⟩ _____ persimmons

**3** Tammy bought 12 pieces of candy, some lollipops for $1.10 each and some chocolate bars for $2.20 each.
She paid $18.70.
**How many of each lollipops and chocolate bars did she buy?**

20 points

⟨Ans.⟩ _____ lollipops
⟨Ans.⟩ _____ chocolate bars

© Kumon Publishing Co., Ltd.    107

6

Date    /    /

Name

■ The Answer Key is on page 184.

## Pattern 4 — Find Two Numbers

● Suppose the sum of the two numbers is 100 and one number is 4 larger than 3 times the other number.
Find these two numbers.

$$\begin{cases} \text{One Number} + \text{The Other Number} = 100 \\ \text{One Number} = \text{The Other Number} \times 3 + 4 \end{cases}$$

## ★ TRY! — Fill in the blanks provided and solve for the answer.

15 points per question

(1) Assume that one number is $x$,

and [        ] is $y$.

(2) And its sum is represented by $x$ and $y$,

[        ]

If one number is 4 larger than 3 times the other number, which is represented by $y$,

[        ]

(3) The simultaneous equations for the magnitude relation of the two numbers are,

$$\begin{cases} \underset{\text{The sum of two numbers}}{[\qquad\qquad]} = 100 \\ x = \underset{\text{The number which is 4 larger than 3 times the othe number}}{[\qquad\qquad]} \end{cases}$$

Solve the above simultaneous equations.

$x =$ [        ] , $y =$ [        ]

⟨Ans.⟩ _____

⟨Ans.⟩ _____

## Hint

In this case, the equation is easier to solve by using the substitution method.

$$\begin{cases} x + y = 100 \\ x = 3y + 4 \end{cases}$$
$$(3y + 4) + y = 100$$

**1** Suppose the sum of the two numbers is 90 and one number is 15 smaller than twice the other number. Fill in the blanks provided and solve for the answer.

15 points

Assume that one number is $x$,

and ⬚ is $y$.

And its sum is represented by $x$ and $y$,

⬚

Represent the number which is 15 smaller than twice the other number, which is represented by $y$,

⬚

The simultaneous equations for the magnitude relation of two numbers are follows,

⬚

Solve the above simultaneous equations.

⟨Ans.⟩ _____

⟨Ans.⟩ _____

**2** Suppose the sum of the two numbers is 140 and one number is 20 smaller than 3 times the other number. Solve the question.

20 points

⟨Ans.⟩ _____

⟨Ans.⟩ _____

**3** Suppose the sum of the two numbers is 210 and one number is 10 larger than 4 times the other number. Solve the question.

20 points

⟨Ans.⟩ _____

⟨Ans.⟩ _____

Date / /

Name

■ The Answer Key is on page 185.

## Pattern 5 — Replacement of Integers

● There is a positive two-digit integer.
That integer is 9 units larger than 4 times the sum of all the digits, and the two-digit number which is created by switching the tens place and the ones place is 18 units larger than the original integer.
Find the original integer.

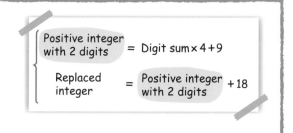

| Positive integer with 2 digits | = Digit sum × 4 + 9 |
| Replaced integer | = Positive integer with 2 digits + 18 |

## ★ TRY! — Fill in the blanks provided and solve for the answer.

15 points per question

(1) Assume that the tens place of the original integer is $x$,

and [          ] is $y$.

(2) If the number which is 9 units larger than 4 times the sum of all the digits is represented by $x$ and $y$,

$4\left([\qquad]\right) + 9$

If the number which is 18 units larger than the original integer is represented by $x$ and $y$,

$\left([\qquad]\right) + 18$

(3) The simultaneous equations for the magnitude relation of the two numbers are,

The numbers which is 9 units larger than 4 times the sum of all the digits

$$10x + y = 4\left([\qquad]\right) + 9$$

The number which is 18 units larger than the original integer

$$10y + x = \left([\qquad]\right) + 18$$

Solve the above simultaneous equations,

$x = [\qquad]$ , $y = [\qquad]$

⟨Ans.⟩ _____

### Hint

Rather than representing the unknown integer as $x$,
the exercise is to assume the tens place is $x$ and the ones place is $y$.

Remember to write a two-digit integer when answering!

**1** **There is a positive two-digit integer.**
**That integer is 1 unit larger than twice the sum of all the digits,**
**and the two-digit number which is created by switching the tens place**
**and the ones place is 54 units larger than the original integer.**
**Fill in the blanks provided and solve for the answer.**

15 points

Assume that the tens place of the original integer is $x$,

and [            ] is $y$.

Because the original integer is 1 unit larger than twice the sum of all the digits,

$$10x + y = 2\left( \boxed{\phantom{xxxx}} \right) + \boxed{\phantom{xxxx}}$$

Since the replaced number is 54 units larger than the original integer,

$$10y + x = \left( \boxed{\phantom{xxxx}} \right) + \boxed{\phantom{xxxx}}$$

Rearrange and solve the above simultaneous equations.

⟨Ans.⟩ _____

**2** **There is a positive two-digit integer.**
**That integer is 7 units larger than twice the sum of all the digits,**
**and the two-digit number which is created by switching the tens place**
**and the ones place is 63 units larger than the original integer.**
**Find the original integer.**

20 points

⟨Ans.⟩ _____

**3** **There is a positive two-digit integer.**
**That integer is 3 units larger than 4 times the sum of all the digits,**
**and the two-digit number created by switching the tens place and the ones**
**place with each other is 27 units larger than the original integer.**
**Find the original integer.**

20 points

⟨Ans.⟩ _____

8

Date / /

Name

■ The Answer Key is on page 185.

## Pattern 6 — Distance During which Speed Changes on the Way

● It took 4 hours to travel 180 km to point C from point A via point B by car.
The speed was 30 km/h between A and B, and 60 km/h between B and C.
Find the distance between A and B, and between B and C respectively.

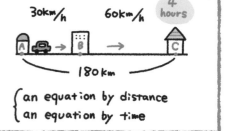

$$\begin{cases} \text{an equation by distance} \\ \text{an equation by time} \end{cases}$$

## ★ TRY! — Fill in the blanks provided and solve for the answer.

15 points per question

( 1 )  Assume that the distance between A and B is $x$ km and ☐ is $y$ km.

( 2 )  If the sum of the distance between A and B, and between B and C is represented by $x$ and $y$,

☐ (km)

If the sum of time required between A and B, and between B and C is represented by $x$ and $y$,

$$\frac{\boxed{\phantom{xx}}}{30} + \frac{\boxed{\phantom{xx}}}{60} \text{ (hours)}$$

( 3 )  The simultaneous equations for the distance and the required time are,

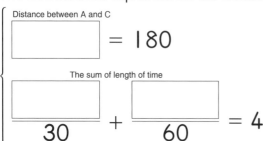

Distance between A and C

$$\boxed{\phantom{xxxx}} = 180$$

The sum of length of time

$$\frac{\boxed{\phantom{xx}}}{30} + \frac{\boxed{\phantom{xx}}}{60} = 4$$

Solve the above simultaneous equations,

$x = \boxed{\phantom{xxx}}$ ,   $y = \boxed{\phantom{xxx}}$

⟨Ans.⟩  Between A and B _____ km

⟨Ans.⟩  Between B and C _____ km

### Hint

"Distance between A and B" + "Distance between B and C" = 180 km.

"Time required between A and B" + "Time required between B and C" = 4 hours.

Also, let's represent by clearing a denominator of fractions in simultaneous equations.

$$\begin{cases} x + y = 180 \\ 2x + y = 240 \end{cases}$$

**1**  It took 4 hours to travel 230 km by car to point C from point A via point B,   15 points
when the speed was 50 km/h between A and B,
and 80 km/h between B and C.
**Fill in the blanks provided and find the distance between
A and B, and between B and C respectively.**

Assume that the distance between A and B is $x$ km

and [＿＿＿＿] is $y$ km.

Because the sum of the distance between is 230 km,

$$\boxed{\phantom{xxxxx}} = 230$$

Since the sum of time between A and B, and between B and C is 4 hours,

$$\boxed{\phantom{xxx}} + \boxed{\phantom{xxx}} = 4$$

Rearrange and solve the above simultaneous equations.

⟨**Ans.**⟩  Between A and B _____ km

⟨**Ans.**⟩  Between B and C _____ km

**2**  It took 3 hours to travel 150 km by car to point C from point A via point B,   20 points
when the speed was 40 km/h between A and B, and 70 km/h between
B and C.
**Find the distance between A and B, and between B and C respectively.**

⟨**Ans.**⟩  Between A and B _____ km

⟨**Ans.**⟩  Between B and C _____ km

**3**  It took James 5 hours to ride 130 miles by motorcycle to point C from   20 points
point A via point B, when his speed was 25 miles per hour between A
and B and 35 miles per hour between B and C.
**Find the distance he rode between A and B, and between B and C respectively.**

⟨**Ans.**⟩  Between A and B _____ miles

⟨**Ans.**⟩  Between B and C _____ miles

113

# Simultaneous Linear Equations 9
## (Pattern 7)

Level ★★

Score
/100

■ The Answer Key is on page 185.

## Pattern 7 — Time that Differs in Going and Returning

● Pass B is located between Town A and Town C, when making a round trip between Town A and Town C.
It took 8 hours to go to Town C at a speed of 30 km/h between A and B, and 50 km/h between B and C.
On the way back, it took 7 hours to return to Town A at a speed of 40 km/h between C and B, and 60 km/h between B and A.
Find the distance between A and B, and between B and C.

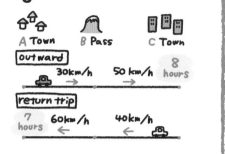

## ★ TRY! — Fill in the blanks provided and solve for the answer.

15 points per question

( 1 ) Assume that the distance between A and B is $x$ km, and [　　　] is $y$ km.

( 2 ) If the sum of time (hours) to go to C is represented by $x$ and $y$,

(Required time between A and B) + (Required time between B and C)

$$\frac{[\quad]}{30} + \frac{[\quad]}{50} \text{ (hours)}$$

If the sum of time (hours) to return to A is represented by $x$ and $y$,

(Required time between B and A) + (Required time between C and B)

$$\frac{[\quad]}{60} + \frac{[\quad]}{40} \text{ (hours)}$$

( 3 ) The simultaneous equations are,

$$\begin{cases} \dfrac{[\quad]}{30} + \dfrac{[\quad]}{50} = 8 \\ \dfrac{[\quad]}{60} + \dfrac{[\quad]}{40} = 7 \end{cases}$$

Solve the above simultaneous equations,

$x = $ [　　　] , $y = $ [　　　]

⟨Ans.⟩ Between A and B　　　km　　⟨Ans.⟩ Between B and C　　　km

### Hint

"Required time between A and B" + "Required time between B and C" = 8 hours

"Required time between B and A" + "Required time between C and B" = 7 hours

"Time" = "Distance" ÷ "Speed"

Let's think about each section.

114　© Kumon Publishing Co., Ltd.

**1** There was Pass B between Town A and Town C, and we made a round trip between A and C by car.

It took 8 hours to go to C at the speed of 40 km/h between A and B and 50 km/h between B and C.

On the way back, it took 9 hours to return to A at the speed of 30 km/h between C and B and 50 km/h between B and A.

Fill in the blanks provided and find the distance between A and B, and between B and C respectively.

25 points

Assume that the distance between A and B is $x$ km

and [_____] is $y$ km.

Because the sum of time (hours) to go to C is 8 hours,

[_____] + [_____] = 8

Since the sum of time to return to A is 9 hours,

[_____] + [_____] = 9

Rearrange and solve the above simultaneous equations.

⟨**Ans.**⟩ Between A and B _____ km

⟨**Ans.**⟩ Between B and C _____ km

**2** Pass B is between Town A and Town C, and we made a round trip between A and C on foot. It took 5 hours to go to Town C at a speed of 3 km/h between Town A and Pass B, and 5 km/h between Pass B and Town C. On the way back, it took 6 hours to return to Town A at a speed of 3 km/h between C and B, and 6 km/h between B and A.

Find the distance between A and B, and between B and C respectively.

30 points

⟨**Ans.**⟩ Between A and B _____ km

⟨**Ans.**⟩ Between B and C _____ km

10

Date    /    /

Name

/100

■ The Answer Key is on page 185.

## Pattern 8 — Passing over an Iron Bridge or Through a Tunnel

● It took a train 100 seconds to pass over an iron bridge that is 2,400 m from beginning to end.
It took the same train 160 seconds to pass through a tunnel that is 3,900 m from beginning to end.
Find the length and the speed of this train in m/sec.

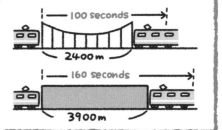

★ **TRY!** — Fill in the blanks provided and solve for the answer.

10 points per question

(1) Assume the length of the train is $x$ m and its speed is $y$ $\boxed{\text{m/} \phantom{xxxx}}$ .

(2) If the distance from the beginning of the bridge to the end is represented by $x$,

$$\boxed{\phantom{xxxx}} + 2400 \text{ (m)}$$

then, if the distance which the train traveled for 100 seconds is represented by $y$,

$$\boxed{\phantom{xxxx}} \text{ (m)}$$

(3) If the distance from the beginning to the end of the tunnel is represented by $x$,

$$\boxed{\phantom{xxxx}} + 3900 \text{ (m)}$$

then, if the distance which the train traveled for 160 seconds is represented by $y$,

$$\boxed{\phantom{xxxx}} \text{ (m)}$$

(4) If the distance is from the start of the tunnel to the end

$$\begin{cases} x + 2400 = \boxed{\phantom{xxxx}} \\ x + 3900 = \boxed{\phantom{xxxx}} \end{cases}$$

Solve the above simultaneous equations.

$$x = \boxed{\phantom{xxxx}} \quad , \quad y = \boxed{\phantom{xxxx}}$$

⟨**Ans.**⟩ Length of the train _____ m     ⟨**Ans.**⟩ Speed (rate) _____ m/sec

### Hint

"The length of the train" + "The length of the iron bridge" = "The distance the train traveled for 100 seconds"

"The length of the train" + "The length of the tunnel" = "The distance the train traveled for 160 seconds"

Do not forget to include the length of the train!

**1** It took a train 65 seconds from starting to cross an iron bridge that was 1,440 m to completely pass over the bridge.
It also took the train 75 seconds from when it entered a tunnel that was 1,680 m to completely pass through the tunnel.
Fill in the blanks provided and find the length and the speed in m/sec of this train.

20 points

Assume the length of the train is $x$ m and its speed is $y$ $\boxed{\text{m} /\phantom{xxx}}$ .

Because it took the train 65 seconds from starting to cross the iron bridge of 1,440 m to completely passing over it,

(Length of the train) + (Length of the iron bridge) = (Distance traveled in 65 seconds)

$$x \;+\; \boxed{\phantom{xxxx}} \;=\; \boxed{\phantom{xxxx}}$$

Since it took the train 75 seconds from starting to enter a tunnel of 1,680 m to completely passing through the tunnel,

(Length of the train) + (Length of the tunnel) = (Distance traveled in 75 seconds)

$$x \;+\; \boxed{\phantom{xxxx}} \;=\; \boxed{\phantom{xxxx}}$$

Rearrange and solve the above simultaneous equations.

⟨**Ans.**⟩ Length _____ m

⟨**Ans.**⟩ Speed (rate) _____ m/sec

**2** It took a train 84 seconds from starting to cross an iron bridge that was 1,900 m to completely pass over the bridge.
It also took the train 104 seconds from starting to enter a tunnel that was 2,400 m to completely pass through the tunnel.
Find the length and the speed in m/sec of this train.

20 points

⟨**Ans.**⟩ Length _____ m

⟨**Ans.**⟩ Speed (rate) _____ m/sec

**3** It took a train 45 seconds from starting to cross an iron bridge that was 1,200 m to completely pass over the bridge.
It took the same train 31 seconds from starting to enter a tunnel that was 780 m to completely pass through the tunnel.
Find the length and speed of the train in m/sec.

20 points

⟨**Ans.**⟩ Length _____ m

⟨**Ans.**⟩ Speed (rate) _____ m/sec

Level

Score

/100

Date / /

Name

■ The Answer Key is on page 185.

## Pattern 9 — Meeting and Catching up

● Around a pond that is 4,800 m, Carl rides a bike and Terry walks.
If they start at the same point at the same time, but go in opposite
directions around the pond, they will meet each other in 16 minutes.
Additionally, if they go around the pond in the same direction,
Carl will catch up to Terry in 24 minutes.
Find the speed in m/min for both Carl and Terry.

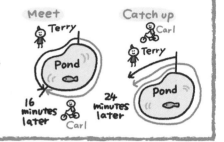

## ★ TRY! — Fill in the blanks provided and solve for the answer.

20 points per question

( 1 ) Assume that Carl's speed is $x$ m/min and Terry's speed is $y$ m/min.
If the sum of the distance Carl and Terry traveled in 16 minutes is represented by $x$ and $y$,

$$16 \boxed{\phantom{xx}} + 16 \boxed{\phantom{xx}} \text{ (m)}$$

If the difference in which the distance Carl traveled minus the distance Terry traveled
in 24 minutes is represented by $x$ and $y$,

$$24 \boxed{\phantom{xx}} - 24 \boxed{\phantom{xx}} \text{ (m)}$$

( 2 ) The simultaneous equations are,

Sum of distance Carl and Terry traveled

$$16 \boxed{\phantom{xx}} + 16 \boxed{\phantom{xx}} = 4800$$

Difference in which distance Carl traveled minus distance Terry traveled

$$24 \boxed{\phantom{xx}} - 24 \boxed{\phantom{xx}} = 4800$$

Solve the above simultaneous equations.

$$x = \boxed{\phantom{xx}} , \quad y = \boxed{\phantom{xx}}$$

⟨Ans.⟩ Carl : Speed (rate) _____ m/min

⟨Ans.⟩ Terry : Speed (rate) _____ m/min

---

**Hint**

When meeting each other,

"Distance A traveled" + "Distance B traveled" = "Distance of a pond per lap"

When catching up,

"Distance A traveled" − "Distance B traveled" = "Distance of a pond per lap"

## Key Point

● When meeting each other
  ⇒ The sum of the distance two people traveled is the distance per lap.
● When catching up
  ⇒ The difference of the distance between two people is the distance per lap.

**1**

**Around a pond that is 2,800 m, Tom rides a bike and Paul walks.**
**If they start at the same point at the same time and go around in opposite directions, they will meet each other in 10 minutes.**
**Additionally, if they go around the pond in the same direction,**
**Tom will catch up to Paul in 20 minutes.**
**Fill in the blanks provided and find the speed in m/min both of Tom and Paul.**

30 points

Assume that Tom's speed is $x$ m/min and Paul's speed is $y$ m/min.

Because the sum of the distance Tom and Paul traveled is equal to 2,800 m of distance per lap of the pond,

(Distance Tom traveled) + (Distance Paul traveled) = (Distance per lap of the pond)

$$\boxed{\phantom{XXXXXXXXXX}} = 2800$$

Since the difference in which distance Tom traveled minus the distance Paul traveled is equal to 2,800 m of the distance around the pond per lap,

(Distance Tom traveled) − (Distance Paul traveled) = (Distance per lap of the pond)

$$\boxed{\phantom{XXXXXXXXXX}} = 2800$$

Rearrange and solve these simultaneous equations.

〈**Ans.**〉 Tom : Speed (rate) _____ m/min

〈**Ans.**〉 Paul : Speed (rate) _____ m/min

**2**

**Around a pond that is 2,100 m, both Stanley and Henry ride their bikes.**
**If they start at the same point at the same time and go around the pond in opposite directions, they meet each other 5 minutes later.**
**Additionally, if they go around the pond in the same direction,**
**Stanley catches up with Henry in 35 minutes.**
**Find the speed in m/min for each person.**

30 points

〈**Ans.**〉 Stanley : Speed (rate) _____ m/min

〈**Ans.**〉 Henry : Speed (rate) _____ m/min

■ The Answer Key is on page 186.

## Pattern 10 — Percentage

● There are 230 male and female students in the seventh grade at a junior high school. Of that, 40% of male students and 55% of female students are in culture clubs, and the total number of students in culture clubs is 110 people. Find the number of male and female students in culture clubs in the seventh grade of this junior high school.

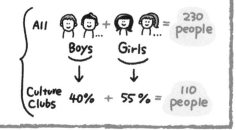

★ **TRY!** — Fill in the blanks provided and solve for the answer.

20 points per question

(1) Assume that the number of male students is $x$ people and the number of female students is $y$ people. If the total number of male and female students is represented by $x$ and $y$,

$$\boxed{\phantom{xxx}} + \boxed{\phantom{xxx}} \text{ (people)}$$

If the total number of 40% of male student and 55% of female students is represented by $y$,

40% of $x$ people $\Rightarrow x \times \dfrac{40}{100}$ people      55% of $y$ people $\Rightarrow y \times \dfrac{55}{100}$ people

$$\dfrac{\boxed{\phantom{xxx}}}{100}x + \dfrac{\boxed{\phantom{xxx}}}{100}y \text{ (people)}$$

(2) The simultaneous equations are,

$$\begin{cases} \boxed{\phantom{xxx}} + \boxed{\phantom{xxx}} = 230 \\ \dfrac{\boxed{\phantom{xxx}}}{100}x + \dfrac{\boxed{\phantom{xxx}}}{100}y = 110 \end{cases}$$

Solve the above simultaneous equations,

$$x = \boxed{\phantom{xxx}}, \quad y = \boxed{\phantom{xxx}}$$

⟨**Ans.**⟩ Male students _____

⟨**Ans.**⟩ Female students _____

---

### Hint

$$\begin{cases} \text{"The number of male students"} + \text{"The number of female students"} = 230 \text{ people} \\ \text{"40\% of the number of male students"} + \text{"55\% of the number of female students"} = 110 \text{ people} \end{cases}$$

## Key Point

$p$ % of $a$ people $\Rightarrow$ $a \times \dfrac{p}{100}$ people

**1** There are 190 male and female students in the seventh grade at a junior high school. Of that, 50% of male students and 45% of female students are in athletic clubs, and the total number of students in athletic clubs is 90 students. Fill in the blanks provided and find the number of male and female students in the seventh grade of this junior high school who are in athletic clubs.

30 points

Assume the number of male students is $x$ people and the number of female students is $y$ people.

Because the total number of male and female students is 190 people,

$$\boxed{\phantom{xxxx}} + \boxed{\phantom{xxxx}} = 190$$

Since the total number of 50% of male students and 45% of female students in athletic clubs is 90 people,

$$\dfrac{\boxed{\phantom{xxx}}}{100}x + \dfrac{\boxed{\phantom{xxx}}}{100}y = 90$$

Rearrange and solve the above simultaneous equations.

⟨**Ans.**⟩  Male students _____

⟨**Ans.**⟩  Female students _____

**2** There are 280 male and female students in the seventh grade at a junior high school.
Of that, 15% of male students and 25% of female students have taken part is activities to clean a local park, and the total number of these students is 56. Find the number of male and female students in the seventh grade of this junior high school who participated in activities to clean the local park.

30 points

⟨**Ans.**⟩  Male students _____

⟨**Ans.**⟩  Female students _____

# Simultaneous Linear Equations 13 (Pattern 11)

13

Level ★

Score

/100

Date    /    /

Name

■ The Answer Key is on page 186.

## Pattern 11 — Increase and Decrease in the Number of People

● The number of volleyball club members in a high school last year was 35 boys and girls altogether.
This year, the boys increased by 10% compared to last year and the girls also increased by 20%, so the total became 40 people.
Find the numbers of the male and female members last year.

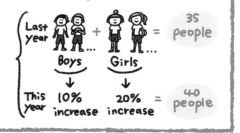

★ **TRY!** — **Fill in the blanks provided and solve for the answer.**

20 points per question

(1) Assume the number of male members was $x$ people and the number of female members was $y$ people last year.
If the total number of male and female members last year is represented by $x$ and $y$,

$$\boxed{\phantom{xxx}} + \boxed{\phantom{xxx}} \text{ (people)}$$

If the total number of male and female members is represented by $x$ and $y$,

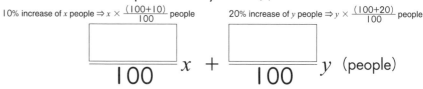

10% increase of $x$ people $\Rightarrow x \times \dfrac{(100+10)}{100}$ people          20% increase of $y$ people $\Rightarrow y \times \dfrac{(100+20)}{100}$ people

$$\dfrac{\boxed{\phantom{xxx}}}{100}x + \dfrac{\boxed{\phantom{xxx}}}{100}y \text{ (people)}$$

(2) The simultaneous equations are,

$$\begin{cases} \boxed{\phantom{xx}} + \boxed{\phantom{xx}} = 35 \\ \dfrac{\boxed{\phantom{xx}}}{100}x + \dfrac{\boxed{\phantom{xx}}}{100}y = 40 \end{cases}$$

Solve the above simultaneous equations.

$$x = \boxed{\phantom{xxx}} , \quad y = \boxed{\phantom{xxx}}$$

⟨**Ans.**⟩   Male students _____

⟨**Ans.**⟩   Female students _____

### Hint

$\begin{cases}$ "The number of male members last year" + "The number of female members last year" = 35 people
"The number of male members this year" + "The number of female members this year" = 40 people

Create equations by "last year's number" and "this year's number" respectively.

## Key Point

$p$ % increase of $a$ people $\Rightarrow a \times \dfrac{(100+p)}{100}$ people

$p$ % decrease of $a$ people $\Rightarrow a \times \dfrac{(100-p)}{100}$ people

**1** **The number of visitors to a museum yesterday was 800 adults and children.** **Today, adults increased by 20% and children decreased by 30% compared to yesterday, so that the number of adults and children totaled 860 people. Fill in the blanks provided and find the number of adults and children who visited the museum yesterday.**  30 points

Assume the number of visitors for adults is $x$ people and the number of visitors for children is $y$ people.

Because the total number of visitors for adults and children was 800 people,

$$\boxed{\phantom{xxxx}} + \boxed{\phantom{xxxx}} = 800$$

Since the total number of visitors for adults and children today is 860 people,

$$\dfrac{\boxed{\phantom{xxxx}}}{100}x + \dfrac{\boxed{\phantom{xxxx}}}{100}y = 860$$

Rearrange and solve the above simultaneous equations.

⟨**Ans.**⟩ Adults _____

⟨**Ans.**⟩ Children _____

**2** **The number of students in a junior high school last year was 600 people in total for boys and girls.** **This year, the number of male students increased by 20% and female students decreased by 10% compared to last year, so the number of boys and girls increased by 15 people in total. Find the numbers of boys and girls in last year's class, respectively.**  30 points

⟨**Ans.**⟩ Male students _____

⟨**Ans.**⟩ Female students _____

Date　　/　　/

Name

■ The Answer Key is on page 186.

## Pattern 12 — Discount

● A shirt and a pair of trousers were bought at a store.
The total list price for both items was $60, but the shirt was
30% off the list price and the trousers were 20% off,
so the final price for the items was $46.
Find the list price of the shirt and trousers, respectively.

Shirt + Trousers = $60

30% OFF + 20% OFF = $46

## ★ TRY! — Fill in the blanks provided and solve for the answer.

20 points per question

(1) Assume the list price of the shirt is $x$ and the list price of the trousers is $y$.
If the total list price of the shirt and the trousers is represented by $x$ and $y$,

$$\boxed{\phantom{xxxx}} + \boxed{\phantom{xxxx}} \quad (\$)$$

If the total final price of the shirt and the trousers after discount is represented by $x$ and $y$,

The price with 30% discount of $\$x$
$\Rightarrow x \times \dfrac{(100-30)}{100}$

The price with 20% discount of $\$y$
$\Rightarrow y \times \dfrac{(100-20)}{100}$

$$\dfrac{\boxed{\phantom{xxxx}}}{100}x + \dfrac{\boxed{\phantom{xxxx}}}{100}y \quad (\$)$$

(2) The simultaneous equations are,

$$\begin{cases} \boxed{\phantom{xxxx}} + \boxed{\phantom{xxxx}} = 60 \\[2mm] \dfrac{\boxed{\phantom{xxxx}}}{100}x + \dfrac{\boxed{\phantom{xxxx}}}{100}y = 46 \end{cases}$$

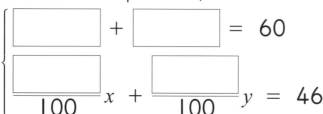
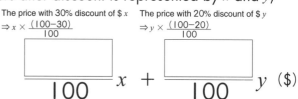

Solve the above simultaneous equations,

$$x = \boxed{\phantom{xxxx}}, \quad y = \boxed{\phantom{xxxx}}$$

⟨**Ans.**⟩　Shirt　　$

⟨**Ans.**⟩　Trousers　$

### Hint

$\begin{cases}\end{cases}$ "The list price of the shirt" + "The list price of the trousers" = $60
"The price of the shirt after discount" + "The price of the trousers after discount" = $46

Let's consider equations with "the list price" and "the price after discount" respectively.

"30% discount" means the price becomes 70% of the list price.

"20% discount" means the price becomes 80% of the list price.

## Key Point

$p$ % discount of $\$a \Rightarrow a \times \dfrac{(100-p)}{100}$ ($\$$)

---

**1** A sweater and a skirt were bought together at a shop. The total list price of the pair was $55, but the sweater was 20% off the list price and the skirt was 10% off the list price, so the final price was $47. Fill in the blanks provided and find the list price of the sweater and skirt, respectively.

20 points

Assume the list price of sweater is $\$x$ and the list price of skirt is $\$y$.

Because the total price of the sweater and skirt is $55.

$$\boxed{\phantom{xxxxx}} + \boxed{\phantom{xxxxx}} = 55$$

Since the total final price of the sweater and skirt is $47,

$$\dfrac{\boxed{\phantom{xxxxx}}}{100}x + \dfrac{\boxed{\phantom{xxxxx}}}{100}y = 47$$

Rearrange and solve the above simultaneous equations.

⟨**Ans.**⟩ Sweater  $ _____

⟨**Ans.**⟩ Skirt  $ _____

---

**2** A suit and tie were bought at a shop. The total list price of the pair was $115, but the suit was 20% off the list price and the tie was 30% off the list price, so the final price became $90. Find the list price of the suit and tie, respectively.

20 points

⟨**Ans.**⟩ Suit  $ _____

⟨**Ans.**⟩ Tie  $ _____

---

**3** Courtney bought a dress and a hat at a store. The total list price for both was $78, but the dress was 25% off and the hat was 50% off, so the final price for the pair was $54. Find the list price of the dress and hat, respectively.

20 points

⟨**Ans.**⟩ Dress  $ _____

⟨**Ans.**⟩ Hat  $ _____

© *Kumon Publishing Co., Ltd.*   125

Date / /

Name

■ The Answer Key is on page 186.

---

**Pattern 13 — Saline Solution 1**

● A teacher wants to make 600 g of a 9% salt solution (saline solution) by mixing a 7% salt solution and a 10% salt solution. Find how many grams of the two kinds of saline solutions should be mixed respectively.

---

★ **TRY!** — Fill in the blanks provided and solve for the answer.

20 points per question

( 1 ) Suppose the teacher mixes $x$ g of a 7% salt solution and $y$ g of a 10% salt solution. If quantity of saline solution is represented by $x$ and $y$,

$$\boxed{\phantom{xxx}} + \boxed{\phantom{xxx}} \quad \text{(g)}$$

If the total quantity of salt contained in a saline solution is represented by $x$ and $y$,

$$\frac{\boxed{\phantom{xxx}}}{100}x + \frac{\boxed{\phantom{xxx}}}{100}y \quad \text{(g)}$$

The quantity of salt contained in 600 g of a 9% salt solution,

$$600 \times \frac{\boxed{\phantom{xxx}}}{100} = \boxed{\phantom{xxx}} \quad \text{(g)}$$

( 2 ) The simultaneous equations are,

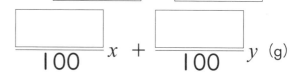

$$\begin{cases} \boxed{\phantom{xxx}} + \boxed{\phantom{xxx}} = 600 \\ \dfrac{\boxed{\phantom{xxx}}}{100}x + \dfrac{\boxed{\phantom{xxx}}}{100}y = \boxed{\phantom{xxx}} \end{cases}$$

Solve the above simultaneous equations,

$$x = \boxed{\phantom{xxx}} \ , \quad y = \boxed{\phantom{xxx}}$$

〈**Ans.**〉 7% salt solution _____ g

〈**Ans.**〉 10% salt solution _____ g

---

**Hint**

⎰ "Quantity of a 7% salt solution" + "Quantity of a 10% salt solution" = 600 g

⎱ "Quantity of salt in a 7% salt solution" + "Quantity of salt in a 10% salt solution"

= "Quantity of salt in 600 g of a 9% salt solution"

Let's create equations about the quantity of a saline solution and the quantity of salt.

---

© Kumon Publishing Co., Ltd.

## Key Point

"Quantity of salt contained in $p$ % salt solution" = "Quantity of saline solution" $\times \dfrac{p}{100}$

**1** A teacher wants to make 700 g of a 10% salt solution by mixing a 5% salt
solution and a 12% salt solution.
**Fill in the blanks provided and find how many grams
of the two kinds of saline solutions should be mixed respectively.**

20 points

Suppose the teacher mixes $x$ g of 5% salt solution and $y$ g of 12% salt solution.

From the relationship of the quantity of a saline solution,

$$\boxed{\phantom{xxxx}} + \boxed{\phantom{xxxx}} = 700$$

From the relationship of the quantity of salt contained in a saline solution,

$$\dfrac{\boxed{\phantom{xxx}}}{100}x + \dfrac{\boxed{\phantom{xxx}}}{100}y = 700 \times \dfrac{\boxed{\phantom{xxx}}}{100}$$

Rearrange and solve the above simultaneous equations.

⟨**Ans.**⟩  5% salt solution _____ g

⟨**Ans.**⟩  12% salt solution _____ g

**2** A teacher wants to make 800 g of a 12% salt solution by mixing a 8% salt
solution and a 13% salt solution.
**Find how many grams of the two kinds of saline solutions should be mixed
respectively.**

20 points

⟨**Ans.**⟩  8% salt solution _____ g

⟨**Ans.**⟩  13% salt solution _____ g

**3** Sally wants to make 350 g of a 22% salt solution by mixing a 18% salt
solution and a 25% salt solution.
**Find how many grams of the two kinds of saline solutions she should mix
together respectively.**

20 points

⟨**Ans.**⟩  18% salt solution _____ g

⟨**Ans.**⟩  25% salt solution _____ g

# Simultaneous Linear Equations 16 (Pattern 14)

Level ★★

Date / /

Name

Score

/100

■ The Answer Key is on page 186.

## Pattern 14 — Saline Solution 2

● A 10% salt solution was created by mixing 300 g of a 12% salt solution with some quantity of a 7% salt solution in grams. Find how many grams of a 7% salt solution was mixed and how many grams of a 10% salt solution was created.

★ **TRY!** — Fill in the blanks provided and solve for the answer.

20 points per question

(1) Assume the mixed 7% salt solution is $x$ g and the produced 10% salt solution is $y$ g. The quantity of salt contained in a 12% salt solution is,

$$300 \times \frac{\boxed{\phantom{00}}}{100} \text{ (g)}$$

The quantity of salt contained in the mixed 7% salt solution is,

$$x \times \frac{\boxed{\phantom{00}}}{100} \text{ (g)}$$

The quantity of salt contained in the mixed 10% salt solution is,

$$y \times \frac{\boxed{\phantom{00}}}{100} \text{ (g)}$$

(2) The simultaneous equations are,

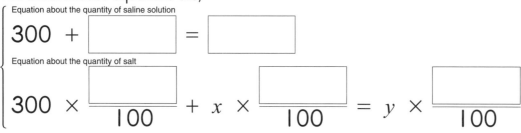

Equation about the quantity of saline solution

$$300 + \boxed{\phantom{0000}} = \boxed{\phantom{0000}}$$

Equation about the quantity of salt

$$300 \times \frac{\boxed{\phantom{00}}}{100} + x \times \frac{\boxed{\phantom{00}}}{100} = y \times \frac{\boxed{\phantom{00}}}{100}$$

Solve the above simultaneous equations,

$$x = \boxed{\phantom{0000}}, \quad y = \boxed{\phantom{0000}}$$

⟨**Ans.**⟩ 7% salt solution _____ g

⟨**Ans.**⟩ 10% salt solution _____ g

### Hint

"Quantity of a 12% salt solution" + "Quantity of a 7% salt solution" = "Quantity of a 10% salt solution"

"Quantity of salt in a 12% salt solution" + "Quantity of salt in a 7% salt solution" = "Quantity of salt in a 10% salt solution"

{ The relationship of the quantity of a saline solution
{ The relationship of the quantity of salt

Create simultaneous equations with these two equations.

**1** A 7% salt solution was created by mixing 200 g of a 10% salt solution with   20 points
some quantity of a 6% salt solution in grams.
**Fill in the blanks provided and find how many grams**
**of a 6% salt solution was mixed and how many grams of a 7% salt solution**
**was created.**

Assume the mixed 6% salt solution is $x$ g and the produced 7% salt solution is $y$ g.

From the relationship of the quantity of a saline solution,

$$200 \; + \; \boxed{\phantom{xxxx}} \; = \; \boxed{\phantom{xxxx}}$$

From the relationship of the quantity of salt contained in a saline solution,

$$200 \times \frac{\boxed{\phantom{xxx}}}{100} \; + \; x \times \frac{\boxed{\phantom{xxx}}}{100} \; = \; y \times \frac{\boxed{\phantom{xxx}}}{100}$$

Rearrange and solve the above simultaneous equations.

⟨**Ans.**⟩   6% salt solution _____ g

⟨**Ans.**⟩   7% salt solution _____ g

**2** A 12% salt solution was created by mixing 600 g of a 15% salt solution   20 points
with some quantity of a 10% salt solution in grams.
**Find how many grams of a 10% salt solution was mixed and how**
**many grams of a 12% salt solution was created.**

⟨**Ans.**⟩   10% salt solution _____ g

⟨**Ans.**⟩   12% salt solution _____ g

**3** Toby created a 15% salt solution by mixing 700 g of a 13% salt solution   20 points
with some quantity of a 17% salt solution in grams.
**Find how many grams of a 17% salt solution was mixed and how many**
**grams of a 15% salt solution he created.**

⟨**Ans.**⟩   17% salt solution _____ g

⟨**Ans.**⟩   15% salt solution _____ g

# Simultaneous Linear Equations Review 1 ★★★

Level

Score

/100

**Date**     /     /

**Name**

■ The Answer Key is on page 187.

① The sum of the two numbers *x* and *y* is 13,
and the sum of 3 times *x* and 4 times *y* is 44.
Find the two numbers *x* and *y*.

10 points

⟨Ans.⟩  *x* = _____

⟨Ans.⟩  *y* = _____

② Deborah bought 15 pieces of fruit.
The apples were $1.40 each and oranges were $0.90 each.
She paid $16.50.
How many apples and oranges did she buy, respectively?

10 points

⟨Ans.⟩ _____ apples

⟨Ans.⟩ _____ oranges

③ Edward bought 18 notebooks in total.
Some notebooks cost $1.20 and the other cost $0.80.
He paid $16.80 for all the notebooks.
How many of the $1.20 and $0.80 notebooks did he buy, respectively?

10 points

⟨Ans.⟩  $1.20 _____ notebooks

⟨Ans.⟩  $0.80 _____ notebooks

④ Stephanie bought 16 flower bulbs.
She bought tulip bulbs for $0.80 each and daffodil bulbs for $0.50 each.
She paid $10.70 total.
How many tulip and daffodil bulbs did Stephanie buy, respectively?

10 points

⟨Ans.⟩ _____ tulip bulbs

⟨Ans.⟩ _____ daffodil bulbs

130    © Kumon Publishing Co., Ltd.

**5** The sum of the prices for 2 notebooks and 5 pencils is $4.     15 points
Additionally, the sum of the prices for 3 notebooks and 8 pencils is $6.20.
What are the prices of 1 notebook and 1 pencil, respectively?

⟨**Ans.**⟩    Notebook    $ _____

⟨**Ans.**⟩    Pencil    $ _____

**6** The admission fee for a museum is $27 for 6 children and 3 adults,     15 points
and $20.50 for 5 children and 2 adults.
How much are the fees for 1 child and 1 adult, respectively?

⟨**Ans.**⟩    Child    $ _____

⟨**Ans.**⟩    Adult    $ _____

**7** The price of 5 pencils and 3 erasers is $4.75.     15 points
Additionally, the price of 3 pencils and 2 erasers is equal.
What is the price of 1 pencil and 1 eraser?

⟨**Ans.**⟩    Pencil    $ _____

⟨**Ans.**⟩    Eraser    $ _____

**8** The price of 7 oranges and 3 apples is $5.80.     15 points
Additionally, the price of 5 oranges and 2 apples is equal.
What is the price of 1 orange and 1 apple?

⟨**Ans.**⟩    Orange    $ _____

⟨**Ans.**⟩    Apple    $ _____

# 18 Simultaneous Linear Equations Review 2 ★★★

Level ★★★

Date / /

Name

Score

/100

■ The Answer Key is on page 187.

**1** Carol and her younger sister Michelle got $8.40 from their mother. They divided the money so that Carol received $3.40 more than Michelle. How much did Carol and Michelle get, respectively?

10 points

⟨Ans.⟩ Carol    $

⟨Ans.⟩ Michelle    $

**2** Kevin and his younger brother Brian got $8.40 from their father. They divided it so that Kevin's money was twice Brian's money. How much money did Kevin and Brian have, respectively?

15 points

⟨Ans.⟩ Kevin    $

⟨Ans.⟩ Brian    $

**3** In Amanda's class, the number of students with cavities is 6 more than the number of students without cavities.
When $\frac{1}{8}$ of the number of students with cavities was treated, they totaled the same number as the students without cavities.
How many students of the original group of students have cavities and do not have cavities, respectively?

15 points

⟨Ans.⟩ Cavities

⟨Ans.⟩ No cavities

**4** In Melissa's middle school, there are 650 students including both boys and girls.
If $\frac{1}{7}$ of the number of girls was added to $\frac{1}{6}$ of the number of boys, it would be 100 students.
How many boys and girls are in Melissa's middle school?

15 points

⟨Ans.⟩ Boys

⟨Ans.⟩ Girls

© Kumon Publishing Co., Ltd.

**5** There is a train running at a constant speed.
It took 65 seconds for this train to start crossing over an iron bridge
1,100 meters long.
It took 90 seconds for the same train to enter and exit a tunnel with a length of
1,550 m. What is the length of this train and how fast is its speed per hour?
[ Hint : Finally, change the meter per second to kilometer per hour. ]

| ⟨Ans.⟩ | Speed | km/h |
|---|---|---|
| ⟨Ans.⟩ | Length | m |

**6** Andrew walked for 19 km from his town to the next town across the hill.
He walked from his town to the hill at 3 km/h, and walked from the hill
to the next town at 5 km/h, it took 5 hours in total.
Find the time it took for him to walk from his town to the hill and from the
hill to the next town, respectively.

| ⟨Ans.⟩ | Home - Hill | hours |
|---|---|---|
| ⟨Ans.⟩ | Hill - Next town | hours |

**7** A road around a pond is a 6 km long loop. Donna runs around the pond
and Emily walks in the opposite direction. They each start at the same point.
When Donna and Emily depart at the same time, they will meet after 20 minutes.
Additionally, if Donna departs 15 minutes behind Emily, they will meet after 16
minutes. How many meters per minute are Donna and Emily moving forward?
[ Hint : Assume that Donna's speed is $x$, Emily's speed is $y$, and since Emily started after
15 minutes, the equation is expressed as $16x+(15+16)y = 6000$ ]

| ⟨Ans.⟩ | Donna | m/min |
|---|---|---|
| ⟨Ans.⟩ | Emily | m/min |

■ The Answer Key is on page 187.

**1** Paul will go from City A to City C through City B, 18 km away.　10 points
He walked from City A to City B at 3 km/h and from City B to City C at 5 km/h.
As a result, he arrived in 4 hours and 24 minutes.
Find the distance from City A to City B and the distance from City B to
City C, respectively.

⟨Ans.⟩　A - B 　　　　　　km

⟨Ans.⟩　B - C 　　　　　　km

**2** Ashley will drive by car from point A to point C through point B.　10 points
When she drove by car 40 km/h from point A to point B,
and drove from point B to point C at 50 km/h, it took 11 hours in total.
Also, when she drove by car 50 km/h from point A to point B,
and drove from point B to point C at 60 km/h, it took 9 hours in total.
Find the distance from point A to point B and the distance from point B to
point C, respectively.

⟨Ans.⟩　A - B 　　　　　　km

⟨Ans.⟩　B - C 　　　　　　km

**3** Steven went from his home to the library through the park.　10 points
When he walked at a speed of 4 km/h from his home to the park, and
from the park to the library at 6 km/h, it took 5 hours and 25 minutes in total.
Also, when he walked at 6 km/h from his home to the park, and from the park
to the library at 4 km/h, it took 5 hours in total.
Find the distance from his home to the park and the distance from the park
to the library, respectively.

⟨Ans.⟩　Home - Park 　　　　　　km

⟨Ans.⟩　Park - Library 　　　　　　km

**4** A 3% salt solution weighing 100 g represents 100 g of water in which  15 points

3 g of salt is dissolved $\left(= \dfrac{3}{100}\right)$.

What amount of salt would be contained in 100 g of a 14% salt solution?
And how many grams of water are in this same solution?

⟨Ans.⟩  Salt _____ g

⟨Ans.⟩  Water _____ g

**5** A teacher made 400 g of an 8% salt solution by mixing a 10% salt solution  15 points
and a 5% salt solution together.

How many grams of a 10% salt solution and a 5% salt solution should be
mixed to make an 8% salt solution?

⟨Ans.⟩  10% salt solution _____ g

⟨Ans.⟩  5% salt solution _____ g

**6** A science teacher wants to make 800 g of a 15% alcohol solution by  15 points
mixing a 20% alcohol solution and a 4% alcohol solution.

How many grams of each alcohol solution should be mixed together?

⟨Ans.⟩  20% alcohol solution _____ g

⟨Ans.⟩  4% alcohol solution _____ g

**7** A teacher wants to make 700 g of a 6% salt solution by mixing a 3%  15 points
salt solution and an 8% salt solution.

How many grams of each solution should be mixed together?

⟨Ans.⟩  3% salt solution _____ g

⟨Ans.⟩  8% salt solution _____ g

# 20

## Simultaneous Linear Equations Review 4 ★★★

Level

Date / /

Name

Score

/100

■ The Answer Key is on page 188.

**1** Birthday cake A costs $13.40 including the box. Birthday cake B is 20% cheaper than A and costs $11 including the same box. What is the cost of the box?

10 points

⟨Ans.⟩ $ _____

**2** The number of students in Tony's middle school last year was 1,000. This year, the number of boys increased by 10% and the number of girls increased by 15% making the total number of students 1,124 people. Find the number of boys and girls in last year's class.

10 points

⟨Ans.⟩ Boys _____

⟨Ans.⟩ Girls _____

**3** During summer school, the travel cost and the accommodation fee combined was $150 per student.
This year, although the accommodation fee has fallen by 5%, the travel cost has risen by 20%, the total cost per student has increased by $10. Find the travel cost and the accommodation fee per student last year.

10 points

⟨Ans.⟩ Travel cost $ _____

⟨Ans.⟩ Accommodation fee $ _____

**4** Last year, there were 45 members of the tennis club at Mark's middle school. This year, although the number of boys increased by 20%, the number of girls decreased by 20%, the total numbers decreased by 1. Find how many boys and girls are in the tennis club this year.

10 points

⟨Ans.⟩ Boys _____

⟨Ans.⟩ Girls _____

**5** There is a two-digit natural number and the sum of the two digits is **11**.  15 points
When the number of tens place and the number of ones place are switched,
it becomes 45 units larger than the original natural number.
Find the original natural number.

⟨Ans.⟩ _____

**6** There is a two-digit natural number.  15 points
This number is 6 times the sum of the two digits and plus 1 unit.
When the number of tens place and the number of ones place are switched,
it becomes 9 units smaller than the original natural number.
Find the original natural number.

⟨Ans.⟩ _____

**7** There is a type of tea that costs $4 per 100 g and another type of tea that  15 points
costs $2.40 per 100 g.
If the two kinds of tea are mixed to create 400 g of tea that costs $3.60 per 100 g.
How many grams of each tea were mixed to create the new tea?

⟨**Ans.**⟩  $4 _____ g

⟨**Ans.**⟩  $2.40 _____ g

**8** Sandra bought 3 roses and 5 carnations in a flower shop and paid $13.50.  15 points
However, the clerk reversed the price of roses and carnations,
and gave her back $0.60. What was the price of 1 rose and 1 carnation?

⟨**Ans.**⟩  Rose  $ _____

⟨**Ans.**⟩  Carnation  $ _____

# Inequalities 1
## (Solving Inequalities)

**21**

Date      /      /

Name

Level ☆

Score          /100

■ The Answer Key is on page 188.

**Don't forget!**

● An inequality is a mathematical sentence that has a symbol such as $>$ or $<$.
Solving an inequality is similar to solving an equation.

The symbol $>$ is read as "is greater than,"
and the symbol $<$ is read as "is less than."

**1**  **Solve each inequality.**

5 points per question

(1)  $x + 4 > 9$
     $x > 9 - 4$
     $x >$

⟨Ans.⟩ _____

(2)  $x - 7 > 3$

⟨Ans.⟩ _____

(3)  $2x + 8 > 20$

⟨Ans.⟩ _____

(4)  $3x + 9 > 12$

⟨Ans.⟩ _____

(5)  $x - 4 < 11$

⟨Ans.⟩ _____

(6)  $9x + 9 < 8$

⟨Ans.⟩ _____

(7)  $-2x + 2 > -6$
     $-2x > -8$
     $x <$

⟨Ans.⟩ _____

(8)  $-6x - 8 > 4$

⟨Ans.⟩ _____

(9)  $-2x + 6 > 20$

⟨Ans.⟩ _____

(10)  $-x + 9 < -5$

⟨Ans.⟩ _____

(11)  $-2x + 3 < -15$

⟨Ans.⟩ _____

(12)  $-6x + 7 < 7$

⟨Ans.⟩ _____

© Kumon Publishing Co., Ltd.

**Don't forget!**

● Inequalities with $\geq$ or $\leq$ are solved the same way as inequalities with $>$ or $<$.
The symbol $\geq$ is read as "is greater than or equal to",
and the symbol $\leq$ is read as "is less than or equal to".

In case of fractions, multiply both sides by the least common multiple (LCM) of the denominators.

**2** **Solve each inequality.**

4 points per question

(1) $-3x - 8 \geq 7$
$$-3x \geq 15$$
$$x \leq$$

⟨Ans.⟩ _____

(2) $5x - 2(x - 3) \leq 18$

⟨Ans.⟩ _____

(3) $-3x - (4 - 7x) \geq -10 + x$

⟨Ans.⟩ _____

(4) $-3x - \left(\dfrac{1}{2} - x\right) \geq -2$

⟨Ans.⟩ _____

(5) $-5x - \left(\dfrac{3}{4}x - 1\right) \leq 3$

⟨Ans.⟩ _____

(6) $\dfrac{x - 5}{3} > \dfrac{x - 2}{4}$
$$4(x - 5) > 3(x - 2)$$
$$4x - 20 > 3x - 6$$
$$x >$$

⟨Ans.⟩ _____

(7) $\dfrac{x + 4}{2} < \dfrac{x - 7}{3}$

⟨Ans.⟩ _____

(8) $\dfrac{2x - 1}{5} \geq \dfrac{4x + 5}{7}$

⟨Ans.⟩ _____

(9) $\dfrac{x + 9}{3} \leq \dfrac{4x - 5}{9}$

⟨Ans.⟩ _____

(10) $\dfrac{3x - 2}{4} \geq \dfrac{9x + 8}{10}$

⟨Ans.⟩ _____

# Inequalities 2
## (Range of $x$)

**22**

Level ★

Date / /

Name

Score /100

■ The Answer Key is on page 188.

## Don't forget!

(1) $\begin{cases} x > 1 & \cdots① \\ x < 3 & \cdots② \end{cases}$

〈Ans.〉 $1 < x < 3$

(2) $\begin{cases} x < 0 & \cdots① \\ x \geq 1 & \cdots② \end{cases}$

〈Ans.〉 No solution

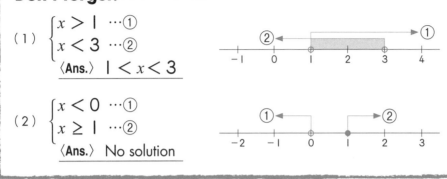

The shaded region shows the range of $x$ that satisfies both inequalities.

---

**1** Draw each inequality on a number line to find the range of $x$ that satisfies both inequalities.

9 points per question

(1) $\begin{cases} x < 5 & \cdots① \\ x > 2 & \cdots② \end{cases}$

〈Ans.〉 $\boxed{\phantom{x}} < x < \boxed{\phantom{x}}$

(2) $\begin{cases} x > 6 & \cdots① \\ x < 0 & \cdots② \end{cases}$

〈Ans.〉 No solution

(3) $\begin{cases} x \geq -1 & \cdots① \\ x < 0 & \cdots② \end{cases}$

〈Ans.〉 $\boxed{\phantom{x}} \leq x < \boxed{\phantom{x}}$

(4) $\begin{cases} x \leq -7 & \cdots① \\ x \geq -10 & \cdots② \end{cases}$

〈Ans.〉 $\boxed{\phantom{x}} \leq x \leq \boxed{\phantom{x}}$

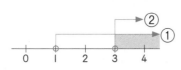
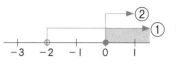
**2** **Find the range of $x$ that satisfies both inequalities.**

8 points per question

( 1 ) $\begin{cases} x > 2 \\ x < 8 \end{cases}$

〈Ans.〉 _____

( 5 ) $\begin{cases} x \leq 5 \\ x < -6 \end{cases}$

〈Ans.〉 _____

( 2 ) $\begin{cases} x > 0 \\ x < -1 \end{cases}$

〈Ans.〉 _____

( 6 ) $\begin{cases} 3x < 2x + 8 & ···① \\ 2x \geq 6 & ···② \end{cases}$

① becomes: $x < \boxed{\phantom{0}}$

② becomes: $x \geq \boxed{\phantom{0}}$

〈Ans.〉 $\boxed{\phantom{0}} \leq x < \boxed{\phantom{0}}$

( 3 ) $\begin{cases} x \geq 2 \\ x \geq -1 \end{cases}$

〈Ans.〉 _____

( 7 ) $\begin{cases} x > 4x - 6 & ···① \\ 5x - 3 > 7x + 9 & ···② \end{cases}$

① becomes:

② becomes:

〈Ans.〉 _____

( 4 ) $\begin{cases} x \leq -2 \\ x > -1 \end{cases}$

〈Ans.〉 _____

( 8 ) $\begin{cases} 6(x - 1) \leq -2(x + 3) & ···① \\ 2 - 6x < -2 + 2x & ···② \end{cases}$

① becomes:

② becomes:

〈Ans.〉 _____

# Inequalities 3
## (Word Problems 1)

**23**

Date / /

Name

Level ★★

Score /100

■ The Answer Key is on page 188.

**1** Write each word problem as an inequality, and then solve.    10 points per question

(1) $x$ plus 6 is greater than four times $x$.

$$x + 6 > \boxed{\phantom{0}} x$$

⟨**Ans.**⟩ _____

(2) $y$ minus 8 is less than or equal to 6.

⟨**Ans.**⟩ _____

(3) 5 times $z$ is greater than or equal to $z$ plus 6.

⟨**Ans.**⟩ _____

(4) $-2$ multiplied by $w$ is less than or equal to 3 less than $w$.

⟨**Ans.**⟩ _____

(5) The sum of 3 and $q$ is greater than 8 times $q$.

⟨**Ans.**⟩ _____

## 2 Write each word problem as an inequality, and then solve.

(1) 3 times $x$ plus 2 is less than 10.

$$\boxed{\phantom{x}}x+2<\boxed{\phantom{x}}$$

⟨Ans.⟩ _____

(2) 3 times the sum of $x$ plus 2 is less than 10.

$$\boxed{\phantom{x}}(x+\boxed{\phantom{x}})<10$$

⟨Ans.⟩ _____

(3) 2 times the sum of $3x$ and 5 is greater than or equal to 9 times $x$.

⟨Ans.⟩ _____

(4) One half the difference of $x$ and 4 is less than twice the difference of 5 and $x$.

$$\frac{1}{2}(x-\boxed{\phantom{x}})<2(\boxed{\phantom{x}}-x)$$

⟨Ans.⟩ _____

(5) The sum of two consecutive integers is less than or equal to 89.
What pair of integers with this property has the greatest sum?
Let the smaller of the two integers be $x$.

⟨Ans.⟩ _____ , _____

# Inequalities 4
## (Word Problems 2)

24

Level ★★

Score

/100

Date / /

Name

■ The Answer Key is on page 189.

**1** **Answer each word problem.**

(1) 10 points per question (2)–(3) 15 points per question

(1) Jack has 50 coins, and Tracey has 20 coins. They each give Patrick the same amount of coins. Jack's new amount of coins is less than or equal to 4 times Tracey's new amount of coins. Find the range of the amount of coins they each could have given Patrick.

Let $x$ be the amount of coins they each gave Patrick.

Jack's new amount of coins is: $50 - x$.

Tracey's new amount of coins is: $20 - \boxed{\phantom{0}}$,

therefore $50 - x \leq \boxed{\phantom{0}} ( 20 - \boxed{\phantom{0}} )$

⟨Ans.⟩ _____ coins or fewer

(2) Maria has 60 toys, and Alex has 40 toys. They each give Jane the same amount of toys. Alex's new amount of toys is greater than or equal to $\frac{1}{2}$ of Maria's new amount of toys. Find the range of the amount of toys they each could have given Jane.

⟨Ans.⟩ _____ toys or fewer

(3) Bobby has 800 books and Logan has 400 books. After they each gave Harry the same number of books, the number of books Bobby has left is less than or equal to 3 times the number of books Logan has left.

Find the range of the number of books they each gave.

⟨Ans.⟩ _____ books or fewer

## 2 Answer each word problem.

20 points per question

(1) Many boys and girls are playing in a park. There are twice as many girls as there are boys. If there are more than 60 children in the park, find the range of the amount of boys that could be in the park.

Let $x$ be the number of boys.

Because there are twice as many girls as boys,

the number of girls is $\boxed{\phantom{0}}x$.

Since there are more than 60 children,

$x + \boxed{\phantom{0}}x > 60$

⟨**Ans.**⟩ More than _____ boys

(2) Bruce sells TVs and radios at his store. The number of radios is half of the number of TVs. If there are at most 285 items in his store, find the range of the amount of radios.

⟨**Ans.**⟩ Less than or equal to _____ radios

(3) There are 200 pencils to be handed out to the students of a class.
If 30 of the students are given 4 each and the others are given 5 each, there will be some pencils left over. However, there are not enough pencils to give 20 of the students 4 each and the others 5 each. Exactly, how many students are there in the class?

⟨**Ans.**⟩ _____ people

# Graphs 1
## (Solving Graphs)

Level ☆

Score

/ 100

Date / /

Name

■ The Answer Key is on page 189.

---

**Don't forget!**

The **y-intercept** of a line is the y-coordinate of the point where the line crosses the y-axis.

For example,
the y-intercept of line ( a ) is −2.
the y-intercept of lines ( b ) and ( c ) is 3.

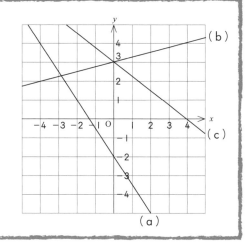

---

**1** **Determine the y-intercept of each line.**

7 points per question

( 1 )   The y-intercept of line ( a ) is [ ].

( 2 )   The y-intercept of line ( b ) is [ ].

( 3 )   The y-intercept of line ( c ) is [ ].

( 4 )   The y-intercept of line ( d ) is [ ].

( 5 )   The y-intercept of line ( e ) is [ ].

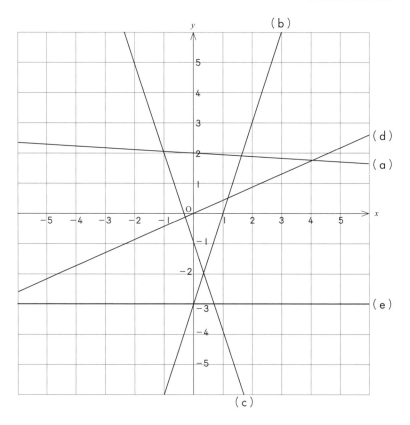

© Kumon Publishing Co., Ltd.

## 2 Determine the *y*-intercept of each equation.

7 points per question

(1)  The *y*-intercept of $y = x + 3$ is ☐.

(2)  The *y*-intercept of $y = 2x - 3$ is ☐.

(3)  The *y*-intercept of $y = -x - 4$ is ☐.

(4)  The *y*-intercept of $y = \dfrac{3}{2}x + 1$ is ☐.

(5)  The *y*-intercept of $y = \dfrac{1}{2}x - 4$ is ☐.

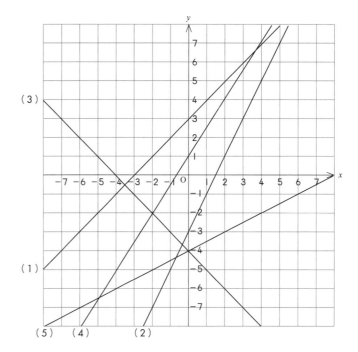

## 3 Determine the *y*-intercept of each equation.

5 points per question

> **Don't forget!**
>
> For example,
> the *y*-intercept of $y = 2x - 5$ is "$-5$".
> the *y*-intercept of $y = \dfrac{1}{2}x + 3$ is "3".

(1)  The *y*-intercept of $y = 5x + 3$ is ☐.

(4)  The *y*-intercept of $y = \dfrac{2}{5}x + \dfrac{3}{4}$ is ☐.

(2)  The *y*-intercept of $y = 2x - 6$ is ☐.

(5)  The *y*-intercept of $y = \dfrac{9}{2}x - \dfrac{13}{8}$ is ☐.

(3)  The *y*-intercept of $y = -x - 9$ is ☐.

(6)  The *y*-intercept of $y = -\dfrac{3}{11}x - \dfrac{25}{13}$ is ☐.

Date / /

Name

■ The Answer Key is on page 189.

**Don't forget!**

The **slope** of a line describes the steepness of a line.

$$\text{slope} = \frac{rise}{run}$$

For example,
the slope of line ( a ) is $\frac{3}{1}$ or 3 because the line rises
3 units vertically for every 1 unit it runs horizontally.

Line ( b ) has a negative slope. The slope of line ( b )
is $-3$ because the line falls 3 units vertically for every
1 unit it runs horizontally.

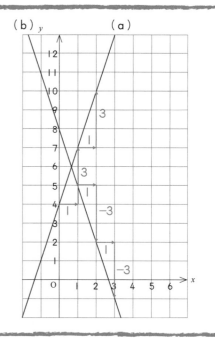

**1** **Graph each equation to find the slope.**

7 points per question

( 1 ) The slope of $y = 2x$ is ☐ .

( 2 ) The slope of $y = 4x$ is ☐ .

( 3 ) The slope of $y = \frac{1}{2}x + 3$ is ☐ .

( 4 ) The slope of $y = \frac{2}{3}x + 3$ is ☐ .

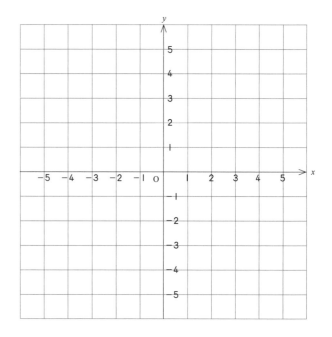

## 2 Graph each equation to find the $y$-intercept.

6 points per question

(1) The $y$-intercept of $y = 2x + 3$ is ☐.

(2) The $y$-intercept of $y = x$ is ☐.

(3) The $y$-intercept of $y = \frac{1}{3}x - 4$ is ☐.

(4) The $y$-intercept of $y = -2x + 1$ is ☐.

(5) The $y$-intercept of $y = -x - \frac{1}{3}$ is ☐.

(6) The $y$-intercept of $y = -\frac{1}{2}x - 3$ is ☐.

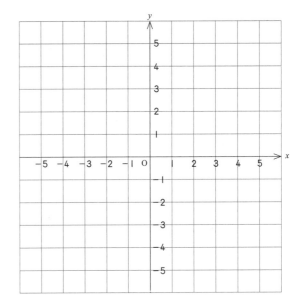

## 3 Determine the slope and $y$-intercept of each equation.

6 points per question

(1) $y = 3x + 9$

⟨**Ans.**⟩ slope _____

⟨**Ans.**⟩ $y$-intercept _____

(2) $y = 4x - \frac{1}{3}$

⟨**Ans.**⟩ slope _____

⟨**Ans.**⟩ $y$-intercept _____

(3) $y = \frac{3}{5}x - \frac{7}{6}$

⟨**Ans.**⟩ slope _____

⟨**Ans.**⟩ $y$-intercept _____

(4) $y = -3x + 2$

⟨**Ans.**⟩ slope _____

⟨**Ans.**⟩ $y$-intercept _____

(5) $y = -\frac{5}{3}x$

⟨**Ans.**⟩ slope _____

⟨**Ans.**⟩ $y$-intercept _____

(6) $y = -7x - \frac{3}{4}$

⟨**Ans.**⟩ slope _____

⟨**Ans.**⟩ $y$-intercept _____

Date    /    /

Name

Level

Score

/100

■ The Answer Key is on page 189.

**1** **Graph each equation to find the slope and *y*-intercept.**

10 points per question

(1) $y = 4x - 1$

〈Ans.〉 slope _____

〈Ans.〉 *y*-intercept _____

(2) $y = x + 3$

〈Ans.〉 slope _____

〈Ans.〉 *y*-intercept _____

(3) $y = -3x + 2$

〈Ans.〉 slope _____

〈Ans.〉 *y*-intercept _____

(4) $y = -x - 3$

〈Ans.〉 slope _____

〈Ans.〉 *y*-intercept _____

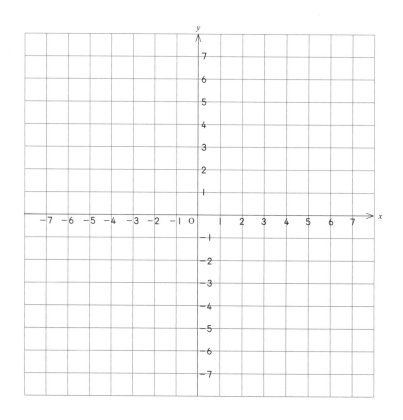

© Kumon Publishing Co., Ltd.

**2** **Graph each equation to find the slope and $y$-intercept.**

(1) $y = \dfrac{3}{2}x - 1$

〈**Ans.**〉 slope _____

〈**Ans.**〉 $y$-intercept _____

(2) $y = -\dfrac{3}{2}x + 1$

〈**Ans.**〉 slope _____

〈**Ans.**〉 $y$-intercept _____

(3) $y = \dfrac{1}{2}x - \dfrac{3}{2}$

〈**Ans.**〉 slope _____

〈**Ans.**〉 $y$-intercept _____

(4) $y = -\dfrac{1}{2}x - \dfrac{3}{2}$

〈**Ans.**〉 slope _____

〈**Ans.**〉 $y$-intercept _____

(5) $y = -\dfrac{2}{3}x - 3$

〈**Ans.**〉 slope _____

〈**Ans.**〉 $y$-intercept _____

(6) $y = -\dfrac{5}{2}x + \dfrac{3}{2}$

〈**Ans.**〉 slope _____

〈**Ans.**〉 $y$-intercept _____

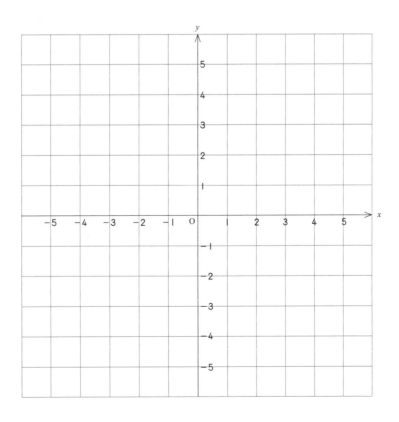

# Graphs 4
## (Word Problems 1)

Date    /    /

Name

Level
★★

Score
/100

■ The Answer Key is on page 189.

**1** Danielle walks at a speed of 2.5 miles per hour.
Use the information to complete each exercise.

12 points per question

(1) Let $x$ represent the number of hours that Danielle walks, and let $y$ represent the distance.
Complete the chart to show the relationship between $x$ and $y$.

| $x$ (hours) | 0 | 1 | 2 | 3 | 4 | 5 |
|---|---|---|---|---|---|---|
| $y$ (miles) | 0 | 2.5 | | | | |

(2) Express the relationship between $x$ and $y$ as an equation.
( Represent in the mixed fraction )

$$y = \boxed{\phantom{xxx}} x$$

⟨Ans.⟩ _____

(3) Graph the equation.

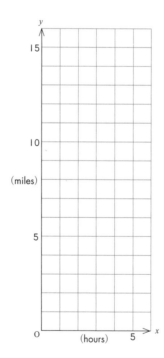

(4) How many miles will Danielle have walked after 6 hours?
Use the graph to answer.

⟨Ans.⟩ _____ miles

**2** **Juan rows a boat at a speed of 15 miles per hour.**
**Use the information to complete each exercise.**

( 1 )  Let $x$ represent the number of hours that Juan rows, and let $y$ represent the distance.
Complete the chart to show the relationship between $x$ and $y$.

| $x$ (hours) | 1 | 2 | 5 | 8 | 10 |
|---|---|---|---|---|---|
| $y$ (miles) |  |  |  |  |  |

( 2 )  Express the relationship between $x$ and $y$ as an equation.

⟨**Ans.**⟩ _____

( 3 )  Graph the equation.

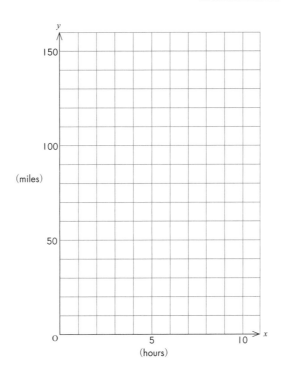

( 4 )  How many miles will Juan have rowed after 9 hours?

$$y = 15 \times \boxed{\phantom{000}} =$$

⟨**Ans.**⟩ _____ miles

153

# Graphs 5
## (Word Problems 2)

29

Date  /  /

Name

Level ★★

Score  /100

■ The Answer Key is on page 190.

**1** **Denise is making biscuits.**
**Use the chart to complete each exercise.**

16 points per question

The chart below shows the relationship between the amount of time she takes ( $x$ minutes ) and the number of biscuits she makes ( $y$ biscuits ).

| $x$ (minutes) | 5 | 10 | 15 | 20 |
|---|---|---|---|---|
| $y$ (biscuits) | 15 | 30 | 45 | 60 |

( 1 )  Express the relationship between $x$ and $y$ as an equation.

⟨**Ans.**⟩ _____

( 2 )  Graph the equation.

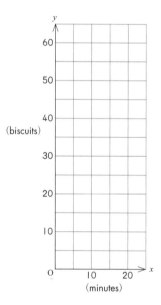

( 3 )  How many biscuits has Denise made after 18 minutes?

⟨**Ans.**⟩ _____  biscuits

**Marilyn is filling a plastic kiddie pool with water. Use the chart to complete each exercise.**

13 points per question

The chart below shows the relationship between the time passed after the water started to be poured ( $x$ minutes ) and the depth ( $y$ cm ) of the water in the pool within that time.

| $x$ (minutes) | 5 | 10 | 15 | 20 | 25 |
|---|---|---|---|---|---|
| $y$ (cm) | 10 | 20 | 30 | 40 | 50 |

( 1 )  Express the relationship between $x$ and $y$ as an equation.

⟨**Ans.**⟩ _____

( 2 )  Graph the equation.

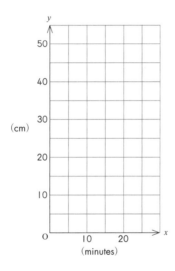

( 3 )  How many centimeters deep is the water in the pool 18 minutes after the water started to be poured?

⟨**Ans.**⟩ _____ cm

( 4 )  How many centimeters does the level of the water go up per minute?

⟨**Ans.**⟩ _____ cm/min

# Graphs 6
## (Word Problems 3)

Date / / Name

Level ★★    Score

/ 100

■ The Answer Key is on page 190.

**1** **Eugene has a machine that makes soap bubbles.**
**Use the chart to complete each exercise.**

15 points per question

The chart below shows the relationship between the amount of time the machine works ( $x$ minutes ) and the number of soap bubbles it makes ( $y$ soap bubbles ).

| $x$ (minutes) | 2 | 3 | 5 | 8 |
|---|---|---|---|---|
| $y$ (soap bubbles) | 14 | 21 | 35 | 56 |

( 1 ) Express the relationship between $x$ and $y$ as an equation.

⟨**Ans.**⟩ _____

( 2 ) Graph the equation.

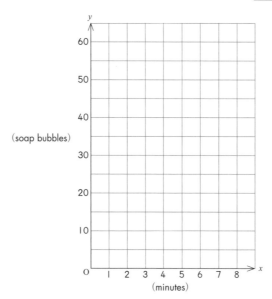

( 3 ) How many soap bubbles are made after 12 minutes?
Use the equation to answer.

⟨**Ans.**⟩ _____ bubbles

( 4 ) How many soap bubbles are made after 21 minutes?
Use the equation to answer.

⟨**Ans.**⟩ _____ bubbles

**2** Amber is walking at a constant speed. Use the chart to complete each exercise.

The chart below shows the relationship between the time passed after she started walking ( $x$ minutes ) and the distance ( $y$ m ) walked within that time.

| $x$ (minutes) | 2 | 5 | 10 | 20 |
|---|---|---|---|---|
| $y$ (m) | 140 | 350 | 700 | 1400 |

(1) Express the relationship between $x$ and $y$ as an equation.

⟨**Ans.**⟩ _____

(2) Graph the equation.

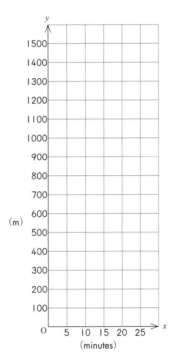

(3) How many times higher is the value of $y$ when $x = 6$ than $x = 3$? Use the equation to answer.

⟨**Ans.**⟩ _____ times

**3** **Answer the question below.**

$y$ varies directly as $x$. When $x = 4$ through the origin $(x, y) = (0, 0)$, the value of $y$ is 12. What is the value of $y$ when $x = 8$?

⟨**Ans.**⟩ _____

31

Date / /

Name

■ The Answer Key is on page 190.

## ★ Check the basics before continuing − practice based questions!

Find words and numbers that apply to ①, ② and ③ in (1), (2) and (3).

### (1) Linear Function

If the relationship between $x$ and $y$ can be expressed as "$y = mx + b$", the equation is called a "linear function".

**★ TRY!**                                                                8 points per question

There is a rectangle with a length of $x$ cm and a width of 5 cm.
Assuming that its circumference ( 2 lengths + 2 widths ) is $y$ cm,
it is expressed as $y = ①x + ②$, and $y$ is a ③ of $x$.

⟨Ans.⟩  ① =

⟨Ans.⟩  ② =

⟨Ans.⟩  ③ =

### (2) Rate of change

The rate of change ( = slope ) of the linear function $y = mx + b$ is constant,
and, it is equal to the coefficient $m$ of $x$.

$$\text{Rate of change} = \frac{\text{Change in } y}{\text{Change in } x}$$

From this equation,
"Change in the amount of $y$" = "Rate of change" × "Change in the amount of $x$" holds.

**★ TRY!**                                                                8 points per question

The rate of change ( = slope ) of the linear function
"$y = 3x + 2$" is ①,
Also, when the increment of $x$ is 4, the increment of $y$ is
"$3 × ② = ③$".

⟨Ans.⟩  ① =

⟨Ans.⟩  ② =

⟨Ans.⟩  ③ =

### (3) Graph of a linear function

The graph of the linear function "$y = mx + b$" is a straight line whose slope is $m$ and $y$-intercept is $b$.

**★ TRY!**                                                                8 points per question

The equation with the slope of 6 and $y$-intercept of 4 is "$y = ①x + ②$".

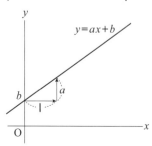

⟨Ans.⟩  ① =

⟨Ans.⟩  ② =

## (4) **How to find an equation for a straight line (Linear Function)**

① When the slope on a graph $m$ and $y$-intercept $b$ are known,

⇒ Substitute into the expression of "$y = mx + b$"

★ **TRY!**   9 points per question

Find the equation of a straight line whose slope on a graph is $\frac{1}{2}$ and whose $y$-intercept is 6.

⟨**Ans.**⟩ _____

② When the slope on a graph ( rate of change ) $m$ and the coordinates of one point ( one set of $x$ and $y$ values ) through which the straight line passes are known,

⇒ Substitute into the expression of "$y = mx + b$" and find the value of $b$.

★ **TRY!**   9 points per question

Find the linear function whose rate of change is $-3$, and $y = 1$ when $x = 3$

⟨**Ans.**⟩ _____

③ When the coordinates of two points through which the straight line passes are known,

⇒ Substitute into the expression of "$y = mx + b$" and find the value of $b$.

★ **TRY!**   9 points per question

Find the equation of a straight line on a graph passing through two points $(3, -1)$ and $(4, 5)$.

⟨**Ans.**⟩ _____

## (5) **Solutions and graphs of simultaneous equations**

The coordinates of the intersection of two straight lines are solutions obtained by solving the equations of two straight lines as simultaneous equations.

★ **TRY!**   9 points per question

Find the coordinates of the intersection of two straight lines using, "$5x + 2y = 9$" and "$3x + 4y = 11$".

Solve the simultaneous equations below.

$$\begin{cases} 5x + 2y = 9 \\ 3x + 4y = 11 \end{cases}$$

⟨**Ans.**⟩ $\left( \quad , \quad \right)$

# Linear Functions 2
## (Solving Linear Functions)

Level ★★

Score

 Date  /  /   Name

/100

■ The Answer Key is on page 190.

**1** **Find the equation of each line.**

12 points per question

( 1 )  A line that has a slope of 3 and a $y$-intercept of 8.

⟨**Ans.**⟩ _____

( 2 )  A line that has a slope of $-3$ and a $y$-intercept of $-5$.

⟨**Ans.**⟩ _____

( 3 )  A line that has a slope of $-1$ and a $y$-intercept of $10$.

⟨**Ans.**⟩ _____

( 4 )  A line that has a slope of $\frac{1}{3}$ and a $y$-intercept of 0.

⟨**Ans.**⟩ _____

---

**Hint**

In the case of #1, because the line has a $y$-intercept of 8,
the line passes through $(0 , 8)$

## 2 Find the equation of each line.

13 points per question

(1) A line that passes through $(2, 5)$ and has a slope of $-1$.

⟨**Ans.**⟩ _____

(2) A line that passes through $(-6, -1)$ and has a slope of $\frac{1}{2}$.

⟨**Ans.**⟩ _____

(3) A line that passes through $(-2, 8)$ and has a slope of $-4$.

⟨**Ans.**⟩ _____

(4) A line that passes through the origin and has a slope of $1$.

⟨**Ans.**⟩ _____

**Hint**

The origin is where the $x$-axis and $y$-axis meet,
therefore the coodinates of the origin are $(0, 0)$.

# Linear Functions 3
## (Solving Linear Functions)

Date
/        /

Name

Level

Score
/ 100

■ The Answer Key is on page 190.

**1**  **Find the equation of each line.**                                    12 points per question

(1)  A line that passes through $(1, 6)$ and $(3, 4)$.

⟨**Ans.**⟩ _____

(2)  A line that passes through $(-1, 5)$ and the origin.

⟨**Ans.**⟩ _____

(3)  A line that passes through $(-1, 5)$ and is parallel to $y = 3x$.

⟨**Ans.**⟩ _____

(4)  A line that passes through $(1, 2)$ and is parallel to $y = -\dfrac{2}{3}x + \dfrac{1}{2}$.

⟨**Ans.**⟩ _____

---

**Hint**

In the case of #3, since the slopes of parallel straight lines are equal,
"$y = 3x + b$" can be expressed.

**2** **Find the point of intersection by solving the simultaneous linear equation algebraically.**

13 points per question

(1) Find the point of intersection of two straight lines
"$3x - 2y = 10$" and "$x - 2y = -6$."

⟨**Ans.**⟩ $(x, y) = ($ ____ , ____ $)$

(2) Find the point of intersection of two straight lines
"$2x - 4y = -14$" and "$12x + 8y = 12$."

⟨**Ans.**⟩ $(x, y) = ($ ____ , ____ $)$

(3) Find the point of intersection of two straight lines
"$3x + y = -1$" and "$-x + y = 3$."

⟨**Ans.**⟩ $(x, y) = ($ ____ , ____ $)$

(4) Find the point of intersection of two straight lines
"$x - 2y = 7$" and "$-3x + y = -6$."

⟨**Ans.**⟩ $(x, y) = ($ ____ , ____ $)$

**Hint**

The point of intersection is a point where two or more lines cross.

You can find the point of intersection by graphing both equations or by solving them algebraically.

# Linear Functions 4
## (Pattern 1)

**34**

Level ★★

Date / /

Name

Score /100

■ The Answer Key is on page 190.

## Pattern 1 — Relationships among Quantities

● A car uses 1 liter of gasoline to run 15 km.
It departed with 60 liters of gasoline in the car.
Suppose the amount of remaining gasoline is $y$ ℓ when the car runs $x$ km.
Represent $y$ with an expression of $x$.

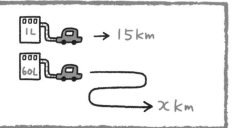

★ **TRY!** — Fill in the blanks provided and solve for the answer.          15 points per question

( 1 )  The amount of gasoline used is proportional to the running [    ] .

If 1 ℓ of gasoline is used to run 15 km,

then [    ] ℓ of gasoline is used to run 1 km.

( 2 )  From #1 above, [    ] ℓ is used to run $x$ km.

( 3 )  "Amount of remaining gasoline" = "Initial amount" − "Amount used to run $x$ km"
Therefore, if $y$ is represented by the expression of $x$,

$$y = \boxed{\phantom{00}} - \boxed{\phantom{00}} x$$

⟨Ans.⟩ _____

---

**Hint**

"Amount of remaining gasoline" = "Initial amount" − "Amount used to run $x$ km"

Once the value of $x$ is decided, the value of $y$ is determined as only one value in the function.

In this question, once the amount of gasoline used to run $x$ km is decided,

the amount of remaining gasoline is determined as only one value, so it is a function.

**1** When a candle with a length of 18 cm was lit, it got shorter    15 points
by 3 cm in 5 minutes.
Let $y$ cm be the length of the candle at $x$ minutes after lighting it.
Fill in the blanks provided, and represent $y$ as a linear function of $x$.

The length of the burning candle is proportional to the time it burns.

Because the candle becomes 3 cm shorter in 5 minutes,

it becomes [        ] cm shorter in 1 minute.

Therefore, it becomes shorter [        ] cm after $x$ minutes.

From this, $y$ is represented by the expression of $x$.

⟨Ans.⟩ _____

**2** When a 12 cm long stick of incense is lit,    20 points
it becames shorter by 6 cm in 5 minutes.
Assume $x$ minutes after the incense is lit, the length of the incense stick is $y$ cm,
and represent $y$ as a linear function of $x$.

⟨Ans.⟩ _____

**3** From the ground up to a certain height, the temperature decreases by 6 °C    20 points
every time the height from the ground increases by 1 km.
When the temperature on the ground is 15 °C, assume the temperature
is $y$ at the point where the height from the ground is $x$ km,
and represent $y$ as a linear function of $x$.

⟨Ans.⟩ _____

# Linear Functions 5
## (Pattern 2)

**35**

Date   /   /

Name

Level ☆☆

Score                    /100

■ The Answer Key is on page 190.

## Pattern 2 — Finding the Length of Spring or the Cost of Making

● The length of a spring when stretched by a certain quantity of hanging weight can be expressed as a linear function.
If a weight of 5 g is hung from a spring, the length of the spring is 25 cm.
If a weight of 8 g is hung from the same spring, the length is 31 cm.
If a weight of 11 g is hung from this spring, how long is the length of the spring?

## ★ TRY! — Fill in the blanks provided and solve for the answer.

20 points per question

(1) Let the length of the initial spring be $b$ cm, and assume the spring increases by $a$ cm each time the hanging weight increases 1 g.
If the length of the spring is $y$ cm when the hanging weight is $x$ g,
the relationship between $x$ and $y$ is,

$$\boxed{\phantom{xxx}} = ax + \boxed{\phantom{xxx}}$$

(2) Because $y = 25$, when $x = 5$,

$$5a + b = \boxed{\phantom{xxx}}$$

Because $y = 31$, when $x = 8$,

$$8a + b = \boxed{\phantom{xxx}}$$

If solving these two simultaneous equations,

$$a = \boxed{\phantom{xxx}}, \quad b = \boxed{\phantom{xxx}}$$

(3) From #2 above, $y = \boxed{\phantom{xxx}} x + \boxed{\phantom{xxx}}$

and if substituting $x = \boxed{\phantom{xxx}}$ in this equation

and find the value of $y$.

⟨Ans.⟩ _____ cm

### Hint

In $y = ax + b$, "$ax$" is proportional and "$+ b$" is the initial length.

**Key Point**

If it is a linear function, you can assume the expression is $y = ax + b$ initially.

**1** The cost of making a photo album can be expressed as a linear function    20 points
in terms of the number of copies produced and the cost.
The cost of making the photo albums is $9,000 for 200 copies and
$12,500 for 300 copies.
Fill in the blanks provided and find the cost making 500 copies of this photo album.

Assume that the cost of making $x$ copies of the photo album is $$y$.

Since $y$ is a [⠀⠀⠀⠀⠀⠀⠀⠀⠀⠀] of $x$, it can be expressed as $y = ax + b$.

When $x = 200$, $y = $ [⠀⠀⠀⠀⠀⠀]

When $x = 300$, $y = $ [⠀⠀⠀⠀⠀⠀]

Then, simultaneous equations can be created and an expression of a linear function can be obtained.

[⠀⠀⠀⠀⠀⠀⠀⠀⠀⠀⠀⠀⠀⠀⠀⠀⠀⠀]

Therefore, find the cost when making 500 copies.

⟨Ans.⟩  $ _____

**2** The cost of making yearbooks for graduates and the number created can    20 points
be represented with a linear function.
The cost of making the yearbooks is $1,900 for 60 items and $2,300 for
75 items.
Find the cost of making 90 yearbooks.

⟨Ans.⟩  $ _____

Date / /

Name

■ The Answer Key is on page 191.

## Pattern 3 — Using Information from a Graph

● Ralph left home, went shopping at a shop on his way, and then went to his aunt's house.

Suppose that he is at a point $x$ minutes after his departure from his house, and the distance from that point to his aunt's house is $y$ km.

The graph on the right shows the relationship between $x$ and $y$.

Find the distance between a point 28 minutes after Ralph left home and his aunt's house.

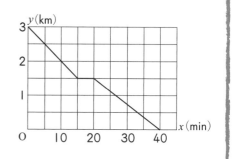

★ **TRY!** — Fill in the blanks provided and solve for the answer.

20 points per question

(1) From the graph, Ralph left the shop how many minutes after he departed from his house?

[ ] minutes

(2) From #1 above,

the graph including his point after 28 minutes is a straight line passing through

the points $\left( [\quad] , \dfrac{3}{2} \right)$ , $( 40 , [\quad] )$ ,

the slope of the line is $-\dfrac{[\quad]}{40}$

Substitute $x = 40$ and $y = 0$ into $\qquad y = [\quad] x + b$ , $b = [\quad]$

(3) Substitute $x = 28$ into the equation obtained in #2 above,

and then, [ ] km. ⟨**Ans.**⟩ _____ km

### Hint

Since the graph that contains the point 28 minutes later is a straight line passing

through the points $\left( 20, \dfrac{3}{2} \right)$, $( 40 , 0 )$, the slope is $\left( 0 - \dfrac{3}{2} \right) \div ( 40 - 20 ) = -\dfrac{3}{40}$.

**Key Point**

Create an equation from the coordinates of two points at the end of the graph that contains the point the question is asking for.

**1** Roy departed from the foot of a mountain, took a break at a mountain cottage on the way, and then walked to the summit. Assuming that $x$ minutes after his departure from the foot, the distance from his point to the summit is $y$ km, the graph on the right shows the relationship between $x$ and $y$. Fill in the blanks provided and find the distance between a point 35 minutes after Roy started from the foot on his way to the summit.

20 points

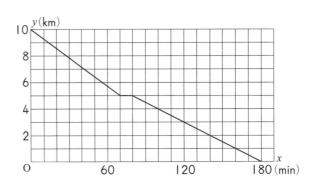

The point 35 minutes after Roy's departure from the foot is before he reaches the mountain cottage. Since the graph is a straight line passing through the points $(0 , 10)$

and $(70 ,$ ⬜ $)$,

the equation representing the relationship between $x$ and $y$ is,

⬜

Find the distance between the point 35 minutes after leaving the foot on his way to the summit.

⟨Ans.⟩ _____ km

**2** Using problem **1** above, find the distance from where Roy departed at the foot of the mountain to the point where Roy walked 150 minutes on his way to the summit.

20 points

⟨Ans.⟩ _____ km

# Linear Functions 7
## (Pattern 4)

**37**

Level ★★

Score

/100

Date / /

Name

■ The Answer Key is on page 191.

## Pattern 4 — Solve Problems Using Information from Two Graphs

● A forest park is 6 km away from Alan and Madison's house.
He walked to the forest park, and she cycled to it.
The figure on the right shows the relationship between time
and distance from their house.
Assuming the distance from their house is $y$ km at $x$ minutes
after 10 o'clock, represent the relationship between $x$ and $y$
for Alan and Madison with respective equations.
Additionally, find the time and distance of the place where
Madison caught up with Alan.

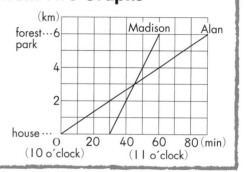

★ **TRY!** — Fill in the blanks provided and solve for the answer.

15 points per question

( 1 ) If the formula for Alan is represented by $y = mx$, the equation of the straight line passing through
the origin and the point $(90 , 6)$ is

$$y = \boxed{\phantom{xxx}}$$

( 2 ) Let the formula for Madison be $y = m'x + b$ and the equation of the straight line
passing through the two points $(30 , 0)$ and $(60 , 6)$ is

$$y = \frac{1}{5}x - \boxed{\phantom{xxx}}$$

( 3 ) The $x$ coordinate of the intersection on the two graphs represents the $\boxed{\phantom{xxx}}$ ,

Madison caught up with Alan, and the $y$ coordinate represents the point where
she caught up with him.

( 4 ) Assume Alan's equation "$y = \frac{1}{15}x$" and Madison's equation "$y = \frac{1}{5}x - 6$"
are simultaneous equations, and solve them.

$$x = \boxed{\phantom{xxx}} , \quad y = \boxed{\phantom{xxx}}$$

( 5 ) From #4 above,

the time when Madison caught up with Alan was $\boxed{\phantom{xxx}}$ minutes after 10 o'clock,

and it is $\boxed{\phantom{xxx}}$ km away from their house.

⟨**Ans.**⟩ Alan's equation _____

⟨**Ans.**⟩ Madison's equation _____

⟨**Ans.**⟩ The time : _____

⟨**Ans.**⟩ The distance _____ km

### Hint

Assuming the intersection for Alan's graph and Madison's graph is P ( $m$ , $n$ ), when Madison caught up with
Alan, the time when she caught up ⇒ The $x$ coordinate $m$ of intersection P
The distance of the place where she caught up ⇒ The $y$ coordinate $n$ of intersection P

**1** **An art museum is 4 km away from Beverly and Wayne's house.** 15 points
**She walked to the art museum, and he cycled to it.**
**The figure on the right shows the relationship between time and distance from their house. Assume the distance from their house is $y$ km at $x$ minutes after 9 o'clock. Fill in the blanks provided and represent the relationship between $x$ and $y$ for Beverly and Wayne with respective equations.**

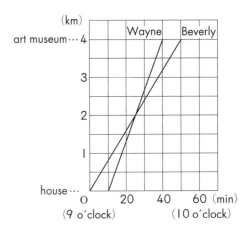

If Beverly's formula is $y = mx$, the slope of the straight line passing

through the point $(50, \boxed{\phantom{xxx}})$,

$$a = \frac{\boxed{\phantom{xxx}}}{\boxed{\phantom{xxx}}} = \boxed{\phantom{xxx}}$$

⟨Ans.⟩ Beverly's equation _____

If Wayne's formula is $y = m'x + b$, the slope of the straight line passing

through the two points $(10, 0)$ and $(40, \boxed{\phantom{xxx}})$,

$$a = \frac{\boxed{\phantom{xxx}} - 0}{\boxed{\phantom{xxx}} - 10} = \boxed{\phantom{xxx}}$$

Substitute $x = 10$ and $y = 0$ into $\quad y = \boxed{\phantom{xxx}} x + b, \quad b = \boxed{\phantom{xxx}}$

⟨Ans.⟩ Wayne's equation _____

**2** **Using problem ① above,** 10 points
**find the time and distance from their house, were Wayne will catch up with Beverly.**

⟨Ans.⟩ Time _____ : _____

⟨Ans.⟩ Distance from their house _____ km

# Linear Functions 8
## (Pattern 5)

**38**

Level ☆☆

Score

/100

Date / /

Name

■The Answer Key is on page 191.

## Pattern 5 — Moving Points

● In rectangle ABCD in the figure on the right, point P starts at A, moves on the side through B and C, and moves to D.
Let the area of △APD be $y$ cm² when point P moves $x$ cm from A.
When point P moves on to and along sides AB, BC and CD, represent the relationship between $x$ and $y$ with an expression.

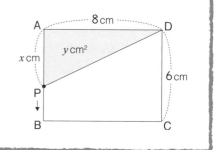

★ **TRY!** — Fill in the blanks provided and solve for the answer.

20 points per question

(1) When point P moves on to and along side AB,

$0 \leq x \leq$ [    ] and hence

$y = \dfrac{1}{2} \times 8 \times$ [    ] = [    ]

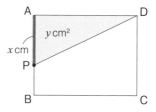

(2) When point P moves on to and along side BC,

$6 \leq x \leq$ [    ] and hence

$y = \dfrac{1}{2} \times 8 \times$ [    ] = [    ]

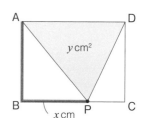

(3) When point P moves on to and along side CD,

$14 \leq x \leq$ [    ] and hence

$y = \dfrac{1}{2} \times 8 \times (6+8+6-$ [    ] $)$

$=$ [    ] $-$ [    ]

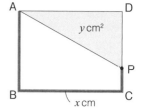

⟨**Ans.**⟩ On to and along side AB

⟨**Ans.**⟩ On to and along side BC

⟨**Ans.**⟩ On to and along side CD

### Hint

In #3 above, the length of PD is ( AB + BC + CD ) − $x$.

**Key Point**

In a question of a moving point, consider it separately for each side.
Look carefully at which part is $x$ cm and represent the length of the side.

**1** In rectangle **ABCD** in the figure on the right, point **P** starts from **B**, moves on to along the sides through **A** and **D**, and moves to **C**. Let the area of △**PBC** be $y$ cm² when point **P** moves $x$ cm from **B**. When point **P** moves on to along sides **BA, AD** and **DC**, fill in the blanks provided and represent the relationship between $x$ and $y$ with an expression.

*20 points*

When point P moves on to along side BA,

$0 \leq x \leq$ [ ] and hence $\quad y = \dfrac{1}{2} \times 7 \times$ [ ] $=$ [ ]

When point P moves on to along side AD,

$4 \leq x \leq$ [ ] and hence $\quad y = \dfrac{1}{2} \times 7 \times$ [ ] $=$ [ ]

When point P moves on to along side DC,

$11 \leq x \leq$ [ ] and hence [ ]

⟨Ans.⟩   On to and along side BA  ————————————

⟨Ans.⟩   On to and along side AD  ————————————

⟨Ans.⟩   On to and along side DC  ————————————

**2** In rectangle **ABCD** in the figure on the right, point **P** starts from **B**, moves on to along the side through **C**, and moves to **D**. Let the area of quadrangle **ABPD** be $y$ cm² when point **P** moves $x$ cm from **B**. When point **P** moves on to along sides **BC** and **CD**, represent the relationship between $x$ and $y$ with an expression.

*20 points*

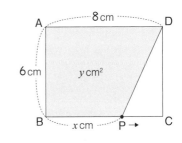

⟨Ans.⟩   On to and along side BC  ————————————

⟨Ans.⟩   On to and along side CD  ————————————

**39** # Linear Functions Review 1 ★★★

Level

Score

/100

Date / /

Name

■ The Answer Key is on page 191.

**1** A phone company offers two different monthly plans for telephone bills, **Plan A and Plan B.** Both plans charge based on a telephone fee, which is determined by the total amount of the basic usage fee plus a call charge which corresponds to the duration of the call. **Answer the following questions.**

8 points per question

| Plan | basic usage fee ( per month ) | call charge ( per minute ) |
|------|------|------|
| A | $16 | $0.30 |
| B | $9.80 | $0.40 |

( 1 ) When the call time for one month is $x$ minutes, the telephone bill is $y$ dollars.
For each of the two plans A and B, show $y$ as an expression of $x$.

⟨**Ans.**⟩   Plan A _____

⟨**Ans.**⟩   Plan B _____

( 2 ) Find how many minutes of call time would make the phone bills for one month the same for both plans A and B.

⟨**Ans.**⟩ _____ minutes

( 3 ) Rebecca is deciding whether to adopt either Plan A or B.
If Rebecca's monthly call time is 50 minutes on average, which plan should she choose?
Write the plan name and reason for her decision.

⟨**Ans.**⟩   Plan    , _____

**2** Ronald departs at 8 am to go to a school that is 1,200 m from his home.
He walked at a speed of 80 m/min.
Now, Ronald is at $y$ m from his school in $x$ minutes after departing from his home.
**Answer the following questions.**

8 points per question

( 1 ) When $x = 5$, find the value of $y$.
( Note that, $y$ is not the distance from his home to the point where Ronald is. )

⟨**Ans.**⟩ _____ m

( 2 ) Find the range of the value of $x$.

⟨**Ans.**⟩ _____

( 3 ) Show $y$ by formula of $x$.

⟨**Ans.**⟩ _____

**3** The figure on the right is rectangle **ABCD** with **AB = 6 cm** and **BC = 8 cm**.

Point **P** starts at point **A** and moves in order of **A → B → C → D** at a speed of 2 cm per second.

Let the area of △**APD** be $y$ cm² when point **P** moves from **A** by taking $x$ seconds.

Answer the following questions.

(1) 8 points (2)–(3) 9 points per question

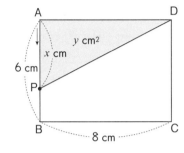

(1) When $x = 2$, find the value of $y$.

〈**Ans.**〉 _____

(2) When $0 \leq x \leq 3$,
Show $y$ by formula of $x$.

〈**Ans.**〉 _____

(3) When $3 \leq x \leq 7$, the value of $y$ is constant.
Find the value of $y$.

〈**Ans.**〉 _____

**4** Laura departed from City A and went to City B which was 14 km away taking a break on the way.

Now, Laura is at $y$ km from City A in $x$ hours after departing from City A.

The figure on the right shows the relationship between $x$ and $y$ at that time.

Answer the following questions.

(1) 8 points (2)–(3) 9 points per question

(1) When $0 \leq x \leq 2$,
Show $y$ by formula of $x$.

〈**Ans.**〉 _____

(2) How long did Laura stop to take a break?

〈**Ans.**〉 _____ hour

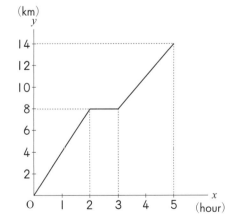

(3) When $3 \leq x \leq 5$,
How fast did Laura walk in kilometers per hour?

〈**Ans.**〉 _____ km/h

# 40 Linear Functions Review 2 ★★★

Level

Score

/100

Date / /

Name

■ The Answer Key is on page 192.

**1** In the figure on the right, there are two points A(3 , 5) and B(12 , 2) on line ℓ, and point C is the intersection of line ℓ and the *x* - axis. Also, point P is on the *x* - axis and its *y* - coordinate is positive. Answer the following questions.

8 points per question

(1) Find the equation of line ℓ.

⟨Ans.⟩

(2) Find the coordinates of point C.

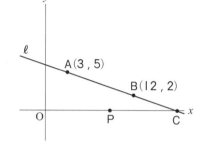

⟨Ans.⟩

(3) When the areas of △AOP and △BPC are equal, find the *x* - coordinate of point P.

⟨Ans.⟩

**2** In the figure on the right, there are three points A(0 , 8) and B(−8 , 0) and C(4 , 0). And point P($k$ , 0) on line OC. Next, take three points Q, R, S on the segments AC, AB, BO, respectively, and make the square PQRS. Answer the following questions.

8 points per question

(1) Find the equation for line AC.

⟨Ans.⟩

(2) Express the *y* - coordinate of point Q with $k$.

⟨Ans.⟩

(3) Find the length of one side of the square PQRS. Use 1 cm as the scale of the coordinates.

⟨Ans.⟩ cm

**3** There are 2 liters of water in aquarium A and 8 liters of water in aquarium B.
Helen started putting water in aquarium A at a rate of 2 ℓ/min.
At the same time, she started draining water from aquarium B at a rate of 1 ℓ/min.
Let $y$ liters be the amount of water in each of aquariums A and B after $x$ minutes, answer the following questions.

9 points per question

(1) Find the equation for the amount of water in aquarium A.

⟨Ans.⟩ _____

(2) Find the equation for the amount of water in aquarium B.

⟨Ans.⟩ _____

(3) Draw graphs to show the equations obtained in #1 and #2, respectively.

(4) How many minutes later will the amount of water in the two tanks be the same?

⟨Ans.⟩ _____ minutes later

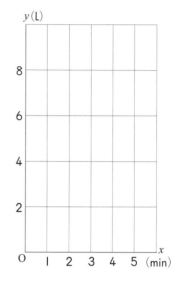

**4** Point P and point Q are separated by 1,200 m.
Timothy goes from point P to point Q. At the same time, Jason goes from point Q to point P.
The figure below shows that they are at $y$ m from point P after $x$ minutes since they departed from their respective points and the relationship between $x$ and $y$ by a graph.
Answer the following questions.

8 points per question

(1) Find how fast Timothy walked in meters per minute.

⟨Ans.⟩ _____ m/min

(2) Find how fast Jason walked in meters per minute.

⟨Ans.⟩ _____ m/min

(3) How long after Timothy and Jason depart, will it take for them to meet?

⟨Ans.⟩ _____ minutes later

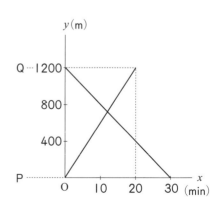

# Review 1

Date / /

Name

■ The Answer Key is on page 192.

**1** Sharon bought 80 stamps.
Several of the stamps cost $0.50 per stamp and several of the stamps
cost $0.80 per stamp.
She paid $50.50 in total.
How many of each type of stamp did Sharon buy?

16 points

〈Ans.〉 $0.50 _____ stamps

〈Ans.〉 $0.80 _____ stamps

**2** There are 12 bags of large and small oranges.
The large bags contains 12 oranges and the small bags contain 9 oranges.
In total there are 120 oranges.
How many of each large and small bags of oranges are there?

16 points

〈Ans.〉 Large _____ bags

〈Ans.〉 Small _____ bags

**3** Jeffrey has $1.20 more than Ryan.
When Ryan hands Jeffrey $7.50,
Jeffrey's money becomes 7 times as much as Ryan's.
How much money did they each have at the beginning?

17 points

〈Ans.〉 Jeffrey $ _____

〈Ans.〉 Ryan $ _____

**4** When 200 g of saline solution A and 400 g of saline solution B are mixed, a 6% salt solution is created.

Also, when 400 g of saline solution A and 200 g of saline solution B are mixed, a 7% salt solution is created.

Find what the percent of concentration of saline solution A and saline solution B are, respectively?

17 points

⟨Ans.⟩  A _____ %

⟨Ans.⟩  B _____ %

**5** At Sandra's middle school, the total number of students last year was 550. This year, the boys decreased by 5% and the girl's increased by 4%, so the total number of students decreased by 5 students.

How many boys and girls were in Sandra's middle school this year?

17 points

⟨**Ans.**⟩  Boys _____

⟨**Ans.**⟩  Girls _____

**6** As part of a beautification activity at Cynthia's junior high school, the students planted flowers in prepared planters.

There are two types of planters, small and large. There are 45 planters in total.

The students planted 2 salvias and 2 marigolds in the small planters, and 7 salvias in the large planter.

The flowers planted in all the prepared planters totaled 231.

How many marigolds were planted in the planters?

[ Hint : Note that two marigolds are planted only in the small planters. ]

17 points

⟨**Ans.**⟩ _____ marigolds

■ The Answer Key is on page 192.

**1** Jacob departed from the starting point of a 10 km running course.
When running at 12 km/h, he felt a pain in his foot,
so he walked at 4 km/h from the middle of the course to the end.
For that reason, it took him 10 minutes longer than when he was running
at 12 km/h to reach the end point.
Find the distance he ran and the distance he walked, respectively.

16 points

[ Hint : Don't forget to add 10 minutes $\left(\frac{10}{60}\right)$ for $\left(\frac{10}{12}\right)$ ]

⟨Ans.⟩ Run _____ km

⟨Ans.⟩ Walk _____ km

**2** There is a post office on the way from Kathleen's home to the library.
It is uphill from Kathleen's home to the post office,
and downhill from the post office to the library.
Kathleen walked to the library and returned home.
She walked uphill at a speed of 80 m/min and downhill at a speed
of 100 m/min.
In total it took her 13 minutes for the journey to the library and 14 minutes
on her way back home.
How long in meters is the distance from Kathleen's home to the post office?

16 points

⟨Ans.⟩ _____ m

**3** Travis purchased 6 pencils and 2 erasers at a fixed price of $4.62.
Since the pencils were half price and the erasers were 40% off the list price,
when Travis bought 8 pencils and 4 erasers it cost $4.20.
Find the list price of 1 pencil and 1 eraser.

17 points

⟨Ans.⟩ Pencil $ _____

⟨Ans.⟩ Eraser $ _____

**4** There is a two-digit natural number.
3 times the number of tens place is 2 units larger than the number of ones place.
Also, twice this natural number is 1 unit larger than the number in which the tens place is switched with the ones place.
Find the original natural number.

17 points

⟨Ans.⟩ _____

**5** Eric and Stephen went fishing on a holiday.
In the morning, Eric caught twice as many fish as Stephen,
and Stephen caught 7 more than Eric in the afternoon.
On this day, Eric caught 23 fish and Stephen caught 24 fish.
Find how many fish did Eric catch in the morning and afternoon, respectively.
[ Hint : Let $x$ be catching in the morning and let $y$ be catching in the afternoon. ]

17 points

⟨Ans.⟩ Morning _____ fish

⟨Ans.⟩ Afternoon _____ fish

**6** The distance from Jonathan's home to the station is 2,800 m.
At first, Jonathan walked at a speed of 80 m/min and ran at a speed of 200 m/min from the midway point.
So, it took him 23 minutes to get to the station after leaving his home.
How many meters did he walk and how many meters did he run, respectively?

17 points

⟨Ans.⟩ Walk _____ m

⟨Ans.⟩ Run _____ m

■ The Answer Key is on page 192.

**1** There are two natural numbers $m$ and $n$.
When $m + n$ is divided by $m$, the quotient is 4 and the remainder becomes 3.
Also, when $10n$ is divided by $m + n$,
the quotient is 7 and the remainder becomes 47.
In this case, find the value of the natural numbers $m$ and $n$.

16 points

⟨Ans.⟩ $m$ _____

⟨Ans.⟩ $n$ _____

**2** At the Sunshine Zoo, there are discount tickets of $2 less per adult and $1 less per child.
If 2 adults and 3 children do not use a discount ticket, the admission fee is $47.
Also, if all 3 adults and 5 children use discount tickets, the admission fee is $63.
Find how much the admission fee is for 1 adult and 1 child without the discount coupons.

16 points

⟨Ans.⟩ Adult $ _____

⟨Ans.⟩ Child $ _____

**3** The distance from Brenda's home to the station is 1,200 m, and there is a bookstore on the way.
She walked from her home to the bookstore at a speed of 60 m/min and from the bookstore to the station at a speed of 80 m/min.
It took her 17 minutes to get from home to the station.
How many meters is the distance from her home to the bookstore and from the bookstore to the station, respectively?

17 points

⟨Ans.⟩ From home to bookstore _____ m

⟨Ans.⟩ From bookstore to station _____ m

**4** In City A, a grant is issued to the Child Association for every 1 kg of recycled material they collect.

17 points

The amount of the grant is different for each recycled material.

The Child Association in District B collected 60 kg of metal and 100 kg of paper, they received $170.

The Child Association in District C collected 40 kg of metal and 150 kg of paper, and received $180.

How much is a grant per 1 kg of metal and how much is a grant per 1 kg of paper?

⟨Ans.⟩ Metal    $ _____

⟨Ans.⟩ Paper    $ _____

**5** A fruit shop purchased 210 oranges to sell.

17 points

They divided the oranges into some bags of 4 and some bags of 6, until the number of bags containing 6 oranges was twice that of the number of bags containing 4 oranges, plus 3 more.

How many bags contain 4 oranges and how many bags contain 6?

⟨Ans.⟩ 4 oranges               bags

⟨Ans.⟩ 6 oranges               bags

**6** The number of students at Anna's middle school was 560 boys and girls 3 years ago.

17 points

Compared to the number of student 3 years ago, this year the number of boys decreased by 18% and the girls increased by 10%.

Then, the total number of students decreased by 5%.

How many boys and girls are there in Anna's class this year?

[ Hint : Note that the answer is the number of boys and girls this year. ]

⟨Ans.⟩ Boys  _____

⟨Ans.⟩ Girls  _____

## 1 Simultaneous Linear Equations 1 (Solving Equations) pp 98, 99

**1** (1) $(x, y) = (3, 4)$  (5) $(x, y) = (0.4, 0.3)$

(2) $(x, y) = (0, \frac{3}{7})$  (6) $(x, y) = (1.1, -0.8)$

(3) $(x, y) = (11, 8)$  (7) $(x, y) = (0.4, -0.3)$

(4) $(x, y) = (2, 0)$  (8) $(x, y) = (0.3, 0.2)$

**2** (1) $(x, y) = (1, -1)$  (5) $(x, y) = (1, 2)$

(2) $(x, y) = (4, 1)$  (6) $(x, y) = (1, -2)$

(3) $(x, y) = (4, 2)$  (7) $(x, y) = (5, 24)$

(4) $(x, y) = (1, 3)$  (8) $(x, y) = (5, 12)$

## 2 Simultaneous Linear Equations 2 (Solving Equations) pp 100, 101

**1** (1) $(x, y) = (3, 8)$  (5) $(x, y) = (3, 2)$

(2) $(x, y) = (3, 5)$  (6) $(x, y) = (-3, 4)$

(3) $(x, y) = (3, 7)$  (7) $(x, y) = (2, 1)$

(4) $(x, y) = (1, -5)$  (8) $(x, y) = (4, 3)$

**2** (1) $(x, y) = (2, 5)$  (5) $(x, y) = (2, 1)$

(2) $(x, y) = (2, 2)$  (6) $(x, y) = (4, -2)$

(3) $(x, y) = (-2, -4)$  (7) $(x, y) = (4, -1)$

(4) $(x, y) = (3, \frac{7}{2})$  (8) $(x, y) = (3, 3)$

## 3 Simultaneous Linear Equations 3 (Pattern 1) pp 102, 103

★ **TRY!**

(1) junior high school student

(2) $2x+y, \ x+2y$

(3) $2x+y, \ x+2y, \ 6, \ 3$  **Ans.** 1 adult $6
**Ans.** 1 student $3

**1** child, $2x+3y, \ x+2y$
$\begin{cases} 2x+3y=26 \\ x+2y=15 \end{cases}$  **Ans.** 1 adult $7
**Ans.** 1 child $4

**2** Let $x$ be the admission fees for an adult, and let $y$ be the fees for a child.
$\begin{cases} 2x+y=26 \\ x+3y=28 \end{cases}$  **Ans.** 1 adult $10
**Ans.** 1 child $6

**3** Let $x$ be the admission fees for an adult, and let $y$ be the fees for a child.
$\begin{cases} 2x+2y=68 \\ x+5y=82 \end{cases}$  **Ans.** 1 adult $22
**Ans.** 1 child $12

## 4 Simultaneous Linear Equations 4 (Pattern 2) pp 104, 105

★ **TRY!**

(1) donut

(2) $5x+4y, \ 6x+3y$

(3) $5x+4y, \ 6x+3y, \ 0.8, \ 0.6$  **Ans.** 1 bread $0.80
**Ans.** 1 donut $0.60

**1** lily, $3x+4y, \ 2x+5y$
$\begin{cases} 3x+4y=13.20 \\ 2x+5y=13 \end{cases}$  **Ans.** 1 rose $2
**Ans.** 1 lily $1.80

**2** Let $x$ be the price of a tulip, and let $y$ be the price of a cosmos.
$\begin{cases} 5x+4y=20.50 \\ 4x+3y=16 \end{cases}$  **Ans.** 1 tulip $2.50
**Ans.** 1 cosmos $2

**3** Let $x$ be the price of a sunflower, and let $y$ be the price of a daffodil.
$\begin{cases} 7x+8y=41.60 \\ 5x+4y=25.60 \end{cases}$  **Ans.** 1 sunflower $3.20
**Ans.** 1 daffodil $2.40

## 5 Simultaneous Linear Equations 5 (Pattern 3) pp 106, 107

★ **TRY!**

(1) puddings

(2) $x+y, \ 1.30x+y$

(3) $x+y, \ 1.30x+y, \ 6, \ 5$  **Ans.** 6 cream puffs
**Ans.** 5 puddings

**1** peaches, $x+y, \ 1.50x+1.80y$
$\begin{cases} x+y=9 \\ 1.5x+1.8y=15 \end{cases}$  **Ans.** 4 apples
**Ans.** 5 peaches

**2** Let $x$ be the number of pears bought by Ralph, and $y$ be the number of persimmons bought by him.
$\begin{cases} x+y=5 \\ 2.40x+1.50y=9.30 \end{cases}$  **Ans.** 2 pears
**Ans.** 3 persimmons

**3** Let $x$ be the number of lollipops bought by Tammy, and $y$ be the number of chocolate bars bought by her.
$\begin{cases} x+y=12 \\ 1.10x+2.20y=18.70 \end{cases}$  **Ans.** 7 lollipops
**Ans.** 5 chocolate bars

## 6 Simultaneous Linear Equations 6 (Pattern 4) pp 108, 109

★ **TRY!**

(1) the other number

(2) $x+y, \ 3y+4$

(3) $x+y, \ 3y+4, \ 76, \ 24$  **Ans.** 76
**Ans.** 24

① the other number, $x+y$, $x=2y-15$
$$\begin{cases} x+y=90 \\ x=2y-15 \end{cases}$$
**Ans.** 55
**Ans.** 35

② Assume that one number is $x$, and another is $y$.
$$\begin{cases} x+y=140 \\ x=3y-20 \end{cases}$$
**Ans.** 100
**Ans.** 40

③ Assume that one number is $x$, and another is $y$.
$$\begin{cases} x+y=210 \\ x=4y+10 \end{cases}$$
**Ans.** 170
**Ans.** 40

## ⑦ Simultaneous Linear Equations 7 (Pattern 5) pp 110, 111

★ **TRY!**
(1) the ones place
(2) $x+y$, $10x+y$
(3) $x+y$, $10x+y$, 5, 7    **Ans.** 57

① the ones place, $x+y$, 1, $10x+y$, 54
   **Ans.** 17

② Assume that the tens digit of the original integer is $x$, and the ones digit is $y$.
$$\begin{cases} 10x+y=2(x+y)+7 \\ 10y+x=(10x+y)+63 \end{cases}$$
**Ans.** 29

③ Assume that the tens digit of the original integer is $x$, and the ones digit is $y$.
$$\begin{cases} 10x+y=4(x+y)+3 \\ 10y+x=(10x+y)+27 \end{cases}$$
**Ans.** 47

## ⑧ Simultaneous Linear Equations 8 (Pattern 6) pp 112, 113

★ **TRY!**
(1) B and C
(2) $x+y$, $x$, $y$
(3) $x+y$, $x$, $y$, 60, 120
   **Ans.** Between A and B 60 km
   **Ans.** Between B and C 120 km

① B and C, $x+y$, $\dfrac{x}{50}$, $\dfrac{y}{80}$
   **Ans.** Between A and B 150 km
   **Ans.** Between B and C 80 km

② Assume that the distance between A and B is $x$ km, and B and C is $y$ km.
$$\begin{cases} x+y=150 \\ \dfrac{x}{40}+\dfrac{y}{70}=3 \end{cases}$$
**Ans.** Between A and B 80 km
**Ans.** Between B and C 70 km

③ Assume that the distance between A and B is $x$ km, and B and C is $y$ km.
$$\begin{cases} x+y=130 \\ \dfrac{x}{25}+\dfrac{y}{35}=5 \end{cases}$$
**Ans.** Between A and B 112.5 miles
**Ans.** Between B and C 17.5 miles

## ⑨ Simultaneous Linear Equations 9 (Pattern 7) pp 114, 115

★ **TRY!**
(1) B and C
(2) $x$, $y$, $x$, $y$
(3) $x$, $y$, $x$, $y$, 120, 200    **Ans.** Between A and B 120 km
   **Ans.** Between B and C 200 km

① B and C, $\dfrac{x}{40}$, $\dfrac{y}{50}$, $\dfrac{x}{50}$, $\dfrac{y}{30}$
   **Ans.** Between A and B 200 km
   **Ans.** Between B and C 150 km

② Assume that the distance between A and B is $x$ km, and B and C is $y$ km.
$$\begin{cases} \dfrac{x}{3}+\dfrac{y}{5}=5 \\ \dfrac{x}{6}+\dfrac{y}{3}=6 \end{cases}$$
**Ans.** Between A and B 6 km
**Ans.** Between B and C 15 km

## ⑩ Simultaneous Linear Equations 10 (Pattern 8) pp 116, 117

★ **TRY!**
(1) m/sec
(2) $x$, $100y$
(3) $x$, $160y$
(4) $100y$, $160y$, 100, 25
   **Ans.** Length of the train 100 m
   **Ans.** Speed 25 m/sec

① m/sec, 1440, $65y$, 1680, $75y$
   **Ans.** Length 120 m
   **Ans.** Speed 24 m/sec

② Assume the length of the train is $x$ m, and its speed is $y$ m/sec.
$$\begin{cases} x+1900=84y \\ y+2400=104y \end{cases}$$
**Ans.** Length 200 m
**Ans.** Speed 25 m/sec

③ Assume the length of the train is $x$ m, and its speed is $y$ m/sec.
$$\begin{cases} x+1200=45y \\ x+780=31y \end{cases}$$
**Ans.** Length 150 m
**Ans.** Speed 30 m/sec

## ⑪ Simultaneous Linear Equations 11 (Pattern 9) pp 118, 119

★ **TRY!**
(1) $x$, $y$, $x$, $y$
(2) $x$, $y$, $x$, $y$, 250, 50    **Ans.** Carl 250 m/min
   **Ans.** Terry 50 m/min

① $10x+10y$, $20x-20y$    **Ans.** Tom 210 m/min
   **Ans.** Paul 70 m/min

② Assume that Stanley's speed is $x$ m/min and Henry's speed is $y$ m/min.
$$\begin{cases} 5x+5y=2100 \\ 35x-35y=2100 \end{cases}$$
**Ans.** Stanley 240 m/min
**Ans.** Henry 180 m/min

**12** **Simultaneous Linear Equations 12 (Pattern 10)** pp 120,121

★ **TRY!**
(1) $x$, $y$, 40, 55
(2) $x$, $y$, 40, 55, 110, 120
    **Ans.** Male students 110
    **Ans.** Female students 120

**1** $x$, $y$, 50, 45     **Ans.** Male students 90
    **Ans.** Female students 100

**2** Assume the number of male students is $x$ people and the number of female students is $y$ people.

$$\begin{cases} x+y=280 \\ \dfrac{15}{100}x+\dfrac{25}{100}y=56 \end{cases}$$

    **Ans.** Male students 140
    **Ans.** Female students 140

---

**13** **Simultaneous Linear Equations 13 (Pattern 11)** pp 122,123

★ **TRY!**
(1) $x$, $y$, 110, 120
(2) $x$, $y$, 110, 120, 20, 15
    **Ans.** Male students 20
    **Ans.** Female students 15

**1** $x$, $y$, 120, 70     **Ans.** Adults 600
    **Ans.** Children 200

**2** Assume the number of male students is $x$ people and the number of female students is $y$ people.

$$\begin{cases} x+y=600 \\ \dfrac{120}{100}x+\dfrac{90}{100}y=615 \end{cases}$$

    **Ans.** Male students 250
    **Ans.** Female students 350

---

**14** **Simultaneous Linear Equations 14 (Pattern 12)** pp 124,125

★ **TRY!**
(1) $x$, $y$, 70, 80
(2) $x$, $y$, 70, 80, 20, 40     **Ans.** Shirt $20
    **Ans.** Trousers $40

**1** $x$, $y$, 80, 90     **Ans.** Sweater $25
    **Ans.** Skirt $30

**2** Let $x$ be the price of a suit, and let $y$ be the price of a tie.

$$\begin{cases} x+y=115 \\ \dfrac{80}{100}x+\dfrac{70}{100}y=90 \end{cases}$$

    **Ans.** Suit $95
    **Ans.** Tie $20

**3** Let $x$ be the price of a dress, and let $y$ be the price of a hat.

$$\begin{cases} x+y=78 \\ \dfrac{75}{100}x+\dfrac{50}{100}y=54 \end{cases}$$

    **Ans.** Dress $60
    **Ans.** Hat $18

---

**15** **Simultaneous Linear Equations 15 (Pattern 13)** pp 126,127

★ **TRY!**
(1) $x$, $y$, 7, 10, 9, 54
(2) $x$, $y$, 7, 10, 54, 200, 400
    **Ans.** 7% salt solution 200 g
    **Ans.** 10% salt solution 400 g

**1** $x$, $y$, 5, 12, 10
    **Ans.** 5% salt solution 200 g
    **Ans.** 12% salt solution 500 g

**2** Suppose the teacher mixes $x$ g of 8% salt solution and $y$ g of 13% salt solution.

$$\begin{cases} x+y=800 \\ \dfrac{8}{100}x+\dfrac{13}{100}y=800\times\dfrac{12}{100} \end{cases}$$

    **Ans.** 8% salt solution 160 g
    **Ans.** 13% salt solution 640 g

**3** Suppose Sally mixes $x$ g of 18% salt solution and $y$ g of 25% salt solution.

$$\begin{cases} x+y=350 \\ \dfrac{18}{100}x+\dfrac{25}{100}y=350\times\dfrac{22}{100} \end{cases}$$

    **Ans.** 18% salt solution 150 g
    **Ans.** 25% salt solution 200 g

---

**16** **Simultaneous Linear Equations 16 (Pattern 14)** pp 128,129

★ **TRY!**
(1) 12, 7, 10
(2) $x$, $y$, 12, 7, 10, 200, 500
    **Ans.** 7% salt solution 200 g
    **Ans.** 10% salt solution 500 g

**1** $x$, $y$, 10, 6, 7
    **Ans.** 6% salt solution 600 g
    **Ans.** 7% salt solution 800 g

**2** Assume the mixed 10% salt solution is $x$ g and the produced 12% salt solution is $y$ g.

$$\begin{cases} 600+x=y \\ 600\times\dfrac{15}{100}+x\times\dfrac{10}{100}=y\times\dfrac{12}{100} \end{cases}$$

    **Ans.** 10% salt solution 900 g
    **Ans.** 12% salt solution 1,500 g

**3** Assume the mixed 15% salt solution is $x$ g and the produced 17% salt solution is $y$ g.

$$\begin{cases} 700+x=y \\ 700\times\dfrac{13}{100}+x\times\dfrac{17}{100}=y\times\dfrac{15}{100} \end{cases}$$

    **Ans.** 17% salt solution 700 g
    **Ans.** 15% salt solution 1,400 g

## (17) Simultaneous Linear Equations Review 1 pp 130, 131

**(1)**
$$\begin{cases} x+y = 13 \\ 3x+4y = 44 \end{cases}$$
**Ans.** $x = 8$
**Ans.** $y = 5$

**(2)** Suppose Deborah bought $x$ apples and $y$ oranges.
$$\begin{cases} x+y = 15 \\ 1.40x+0.90y = 16.50 \end{cases}$$
**Ans.** 6 apples
**Ans.** 9 oranges

**(3)** Assume that Edward bought $x$ notebooks with a price of $1.20 and $y$ notebooks with a price of $0.80.
$$\begin{cases} x+y = 18 \\ 1.20x+0.80y = 16.80 \end{cases}$$
**Ans.** $1.20 6 notebooks
**Ans.** $0.80 12 notebooks

**(4)** Suppose Stephanie bought $x$ tulip bulbs and $y$ daffodil bulbs.
$$\begin{cases} x+y = 16 \\ 0.80x+0.50y = 10.70 \end{cases}$$
**Ans.** 9 tulip bulbs
**Ans.** 7 daffodil bulbs

**(5)** Let $x$ be the price of a notebook, and let $y$ be the price of a pencil.
$$\begin{cases} 2x+5y = 4 \\ 3x+8y = 6.20 \end{cases}$$
**Ans.** Notebook $1
**Ans.** Pencil $0.40

**(6)** Let $x$ be the admission fees for a child, and let $y$ be the fees for an adult.
$$\begin{cases} 6x+3y = 27 \\ 5x+2y = 2.50 \end{cases}$$
**Ans.** Child $2.50
**Ans.** Adult $4

**(7)** Let $x$ be the price of a pencil, and let $y$ be the price of an eraser.
$$\begin{cases} 5x+3y = 4.75 \\ 3x = 2y \end{cases}$$
**Ans.** Pencil $0.50
**Ans.** Eraser $0.75

**(8)** Let $x$ be the price of an orange, and let $y$ be the price of an apple.
$$\begin{cases} 7x+3y = 5.80 \\ 5x = 2y \end{cases}$$
**Ans.** Orange $0.40
**Ans.** Apple $1

## (18) Simultaneous Linear Equations Review 2 pp 132, 133

**(1)** Let $x$ be the amount of money Carol gets and let $y$ be the amount of money Michelle gets.
$$\begin{cases} x+y = 8.40 \\ x-y = 3.40 \end{cases}$$
**Ans.** Carol $5.90
**Ans.** Michelle $2.50

**(2)** Let $x$ be the amount of money Kevin gets and let $y$ be the amount of money Brian gets.
$$\begin{cases} x+y = 8.40 \\ x = 2y \end{cases}$$
**Ans.** Kevin $5.60
**Ans.** Brian $2.80

**(3)** Let $x$ be the number of people in the original group of students that have cavities and let $y$ be the number of people in the original group of students that have no cavities.
$$\begin{cases} x = y+6 \\ x-\frac{1}{8}x = y+\frac{1}{8}x \end{cases}$$
**Ans.** Cavities 24
**Ans.** Not cavities 18

**(4)** Let $x$ be the number of boys and let $y$ be the number of girls.
$$\begin{cases} x+y = 650 \\ \frac{1}{6}x+\frac{1}{7}y = 100 \end{cases}$$
**Ans.** Boys 300
**Ans.** Girls 350

**(5)** Let the speed of this train be $x$ m/sec and the length of this train be $y$ meters.
$$\begin{cases} 65x = 1100+y \\ 90x = 1550+y \end{cases}$$
**Ans.** Speed 64.8 km/h
( 18 m/sec × 60 × 60 ÷ 1000 )
**Ans.** Length 70 m

**(6)** Let $x$ be the time Andrew walked from his town to the hill, and $y$ the time he walked from the hill to the next town.
$$\begin{cases} x+y = 5 \\ 3x+5y = 19 \end{cases}$$
**Ans.** Home - Hill 3 hours
**Ans.** Hill - Next town 2 hours

**(7)** Assumed that the Donna's speed is $x$ m/min and the Emily's speed is $y$ m/min.
$$\begin{cases} 20x+20y = 6000 \\ 16x+(15+16)y = 6000 \end{cases}$$
**Ans.** Donna 220 m/min
**Ans.** Emily 80 m/min

## (19) Simultaneous Linear Equations Review 3 pp 134, 135

**(1)** Let $x$ be the distance from city A to city B, and let $y$ be the distance from city B to city C.
$$\begin{cases} x+y = 18 \\ \frac{x}{3}+\frac{y}{5} = 4\frac{2}{5} \end{cases}$$
**Ans.** A - B 6 km
**Ans.** B - C 12 km

**(2)** Let $x$ be the distance from point A to B, and let $y$ be the distance from point B to C.
$$\begin{cases} \frac{x}{40}+\frac{y}{50} = 11 \\ \frac{x}{50}+\frac{y}{60} = 9 \end{cases}$$
**Ans.** A - B 200 km
**Ans.** B - C 300 km

**(3)** Let $x$ be the distance from Steven's home to the park, and let $y$ be the distance from the park to the library.
$$\begin{cases} \frac{x}{4}+\frac{y}{6} = 5\frac{5}{12} \\ \frac{x}{6}+\frac{y}{4} = 5 \end{cases}$$
**Ans.** Home - Park 15 km
**Ans.** Park - Library 10 km

**(4)** salt $\cdots 100 \times \frac{14}{100} = 14$

water $\cdots 100 - 14 = 86$

**Ans.** Salt 14 g
**Ans.** Water 86 g

(5) Suppose the teacher mixes $x$ g of 10% salt solution and $y$ g of 5% salt solution.

$$\begin{cases} x+y=400 \\ \dfrac{10}{100}x+\dfrac{5}{100}y=400\times\dfrac{8}{100} \end{cases}$$

**Ans.** 10% salt solution 240 g

**Ans.** 5% salt solution 160 g

(6) Suppose the teacher mixes $x$ g of 20% alcohol solution and $y$ g of 4% alcohol solution.

$$\begin{cases} x+y=800 \\ \dfrac{20}{100}x+\dfrac{4}{100}y=800\times\dfrac{15}{100} \end{cases}$$

**Ans.** 20% alcohol solution 550 g

**Ans.** 4% alcohol solution 250 g

(7) Suppose the teacher mixes $x$ g of 3% salt solution and $y$ g of 8% salt solution.

$$\begin{cases} x+y=700 \\ \dfrac{3}{100}x+\dfrac{8}{100}y=700\times\dfrac{6}{100} \end{cases}$$

**Ans.** 3% salt solution 280 g

**Ans.** 8% salt solution 420 g

(20) **Simultaneous Linear Equations Review 4**  pp 136,137

(1) Assume the price of box is $\$x$, and the price for cake A is $\$y$.

$$\begin{cases} x+y=13.40 \\ x+0.8y=11 \end{cases}$$  **Ans.** $\$1.40$

(2) Let $x$ be the number of boys in last year's class, and let $y$ be the number of girls in last year's class.

$$\begin{cases} x+y=1000 \\ 1.1x+1.15y=1124 \end{cases}$$  **Ans.** Boys 520

**Ans.** Girls 480

(3) Let $x$ be the travel cost per student last year, and let $y$ be the accommodation fee per student last year.

$$\begin{cases} x+y=150 \\ 0.2x-0.05y=10 \end{cases}$$  **Ans.** Travel cost $\$70$

**Ans.** Accommodation fee $\$80$

(4) Let $x$ be the number of boys last year, and let $y$ be the number of girls last year.

$$\begin{cases} x+y=45 \\ \dfrac{120}{100}x+\dfrac{80}{100}y=44 \end{cases}$$  **Ans.** Boys 20

**Ans.** Girls 25

(5) Assume that the tens digit of the original integer is $x$, and the ones digit is $y$.

$$\begin{cases} x+y=11 \\ 10y+x=10x+y+45 \end{cases}$$  **Ans.** 38

(6) Assume that the tens digit of the original integer is $x$, and the ones digit is $y$.

$$\begin{cases} 10x+y=6(x+y)+1 \\ 10y+x=10x+y-9 \end{cases}$$  **Ans.** 43

(7) Suppose he/she mixes $x$ g of the tea that cost $\$4$ per 100 g and $y$ g of the tea that cost $\$2.40$ per 100 g.

$$\begin{cases} x+y=400 \\ \dfrac{4}{100}x+\dfrac{2.40}{100}y=\dfrac{3.60}{100}\times400 \end{cases}$$

**Ans.** $\$4$  300 g

**Ans.** $\$2.40$  100 g

(8) Assume the price of a rose is $\$x$, and the price of a carnation is $\$y$.

$$\begin{cases} 3x+5y=13.50-0.60 \\ 5x+3y=13.50 \end{cases}$$  **Ans.** Rose $\$1.80$

**Ans.** Carnation $\$1.50$

(21) **Inequalities 1 (Solving Inequalities)**  pp 138,139

(1) (1) $x>5$        (7) $x<4$
(2) $x>10$       (8) $x<-2$
(3) $x>6$        (9) $x<-7$
(4) $x>1$        (10) $x>14$
(5) $x<15$       (11) $x>9$
(6) $x<-\dfrac{1}{9}$       (12) $x>0$

(2) (1) $x\le5$       (6) $x>14$
(2) $x\le4$       (7) $x<-26$
(3) $x\ge-2$      (8) $x\le-\dfrac{16}{3}$
(4) $x\le\dfrac{3}{4}$       (9) $x\ge32$
(5) $x\ge-\dfrac{8}{23}$      (10) $x\le\dfrac{26}{3}$

(22) **Inequalities 2 (Range of $x$)**  pp 140,141

(1) (1) $2<x<5$

(2) No solution

(3) $-1\le x<0$

(4) $-10\le x\le-7$

(2) (1) $2<x<8$       (5) $x<-6$
(2) No solution    (6) 8, 3    **Ans.** $3\le x<8$
(3) $x\ge2$        (7) $x<-6$
(4) No solution    (8) No solution

(23) **Inequalities 3 (Word Problems 1)**  pp 142,143

(1) (1) $x+6>\boxed{4}x$       **Ans.** $x<2$
(2) $y-8\le6$         **Ans.** $y\le14$
(3) $5z\ge z+6$        **Ans.** $z\ge\dfrac{3}{2}$
(4) $-2w\le w-3$       **Ans.** $w\ge1$
(5) $3+q>8q$          **Ans.** $q<\dfrac{3}{7}$

**②** (1) $\boxed{3}x+2<\boxed{10}$  **Ans.** $x<\dfrac{8}{3}$

(2) $\boxed{3}(x+\boxed{2})<10$  **Ans.** $x<\dfrac{4}{3}$

(3) $2(3x+5)\geq9x$  **Ans.** $x\leq\dfrac{10}{3}$

(4) $\dfrac{1}{2}(x-\boxed{4})<2(\boxed{5}-x)$  **Ans.** $x<\dfrac{24}{5}$

(5) $x+(x+1)\leq89$  **Ans.** 44, 45

## (24) Inequalities 4 (Word Problems 2)  pp 144,145

**①** (1) $x$, 4, $x$  **Ans.** 10 coins or fewer

(2) Let $x$ be the number of toys Maria and Alex each gave Jane.

$40-x\geq\dfrac{1}{2}(60-x)$

$x\leq20$  **Ans.** 20 toys or fewer

(3) Let $x$ be the number of books Bobby and Logan each gave Harry.

$800-x\leq3(400-x)$

$x\leq200$  **Ans.** 200 books or fewer

**②** (1) 2, 2  **Ans.** More than 20 boys

(2) Let $x$ be the amount of radios.

$x+2x\leq285$

$x\leq95$  **Ans.** Less than or equal to 95 radios

(3) Let $x$ be the number of students.

$4\times30+5(x-30)<200$

$x<46\cdots$①

$4\times20+5(x-20)>200$

$x>44\cdots$②

From ① and ②,

$44<x<46$  **Ans.** 45 people

## (25) Graphs 1 (Solving Graphs)  pp 146,147

**①** (1) 2  (4) 0
(2) −3  (5) −3
(3) −1
**②** (1) 3  (4) 1
(2) −3  (5) −4
(3) −4
**③** (1) 3  (4) $\dfrac{3}{4}$
(2) −6  (5) $-\dfrac{13}{8}$
(3) −9  (6) $-\dfrac{25}{13}$

## (26) Graphs 2 (Solving Graphs)  pp 148,149

**①** (1) 2
(2) 4
(3) $\dfrac{1}{2}$
(4) $\dfrac{2}{3}$

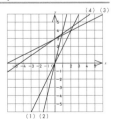

**②** (1) 3
(2) 0
(3) −4
(4) 1
(5) $-\dfrac{1}{3}$
(6) −3

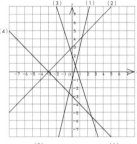

**③** (1) slope 3, $y$-intercept 9

(2) slope 4, $y$-intercept $-\dfrac{1}{3}$

(3) slope $\dfrac{3}{5}$, $y$-intercept $-\dfrac{7}{6}$

(4) slope −3, $y$-intercept 2

(5) slope $-\dfrac{5}{3}$, $y$-intercept 0

(6) slope −7, $y$-intercept $-\dfrac{3}{4}$

## (27) Graphs 3 (Solving Graphs)  pp 150,151

**①** (1) slope 4, $y$-intercept −1
(2) slope 1, $y$-intercept 3
(3) slope −3, $y$-intercept 2
(4) slope −1, $y$-intercept −3

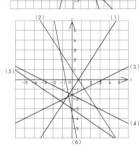

**②** (1) slope $\dfrac{3}{2}$, $y$-intercept −1

(2) slope $-\dfrac{3}{2}$, $y$-intercept 1

(3) slope $\dfrac{1}{2}$, $y$-intercept $-\dfrac{3}{2}$

(4) slope $-\dfrac{1}{2}$, $y$-intercept $-\dfrac{3}{2}$

(5) slope $-\dfrac{2}{3}$, $y$-intercept −3

(6) slope $-\dfrac{5}{2}$, $y$-intercept $\dfrac{3}{2}$

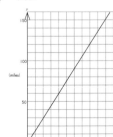

## (28) Graphs 4 (Word Problems 1)  pp 152,153

**①** (1) 5, 7.5, 10, 12.5

(2) $2\dfrac{1}{2}$

**Ans.** $y=2\dfrac{1}{2}x$

(3)

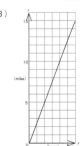

(4) 15 miles

**②** (1) 15, 30, 75, 120, 150
(2) $y=15x$
(3)

(4) $y=15\times\boxed{9}=135$

**Ans.** 135 miles

**㉙ Graphs 5 (Word Problems 2)** pp 154,155

**1** (1) $y = 3x$

(2)
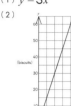

(3) $y = 3 \times 18 = 54$

**Ans.** 54 biscuits

**2** (1) $y = 2x$

(2)
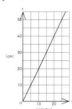

(3) $y = 2 \times 18 = 36$

**Ans.** 36 cm

(4) 2 cm/min

**㉚ Graphs 6 (Word Problems 3)** pp 156,157

**1** (1) $y = 7x$

(2)
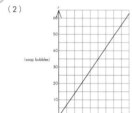

(3) $y = 7 \times 12 = 84$

**Ans.** 84 bubbles

(4) $y = 7 \times 21 = 147$

**Ans.** 147 bubbles

**2** (1) $y = 70x$

(2)
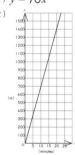

(3) 2 times

**3** 24

**How to Solve**  If $x = 4$ and $y = 12$ are substituted into the equation $y = ax$, then $a = 3$.
Therefore, $y = 3x$.
Substitute $x = 8$ in this equation.
$y = 3 \times 8 = 24$

**㉛ Linear Functions 1 (Solving Linear Functions)** pp 158,159

(1) ① 2, ② 10, ③ linear functions

(2) ① 3, ② 4, ③ 12

(3) ① 6, ② 4

(4) ① $y = \frac{1}{2}x + 6$  ② $y = -3x + 10$  ③ $y = 6x - 19$

(5) (1, 2)

**㉜ Linear Functions 2 (Solving Linear Functions)** pp 160,161

**1** (1) $y = 3x + 8$

(2) $y = -3x - 5$

(3) $y = -x + 10$

(4) $y = \frac{1}{3}x$

**2** (1) $y = -x + 7$

(2) $y = \frac{1}{2}x + 2$

(3) $y = -4x$

(4) $y = x$

**㉝ Linear Functions 3 (Solving Linear Functions)** pp 162,163

**1** (1) $y = -x + 7$

(2) $y = -5x$

(3) $y = 3x + 8$

(4) $y = -\frac{2}{3}x + \frac{8}{3}$

**2** (1) $(x, y) = (8, 7)$

(2) $(x, y) = (-1, 3)$

(3) $(x, y) = (-1, 2)$

(4) $(x, y) = (1, -3)$

**㉞ Linear Functions 4 (Pattern 1)** pp 164,165

**★ TRY!**

(1) distance, $\frac{1}{15}$

(2) $\frac{1}{15}x$

(3) 60, $\frac{1}{15}$

**Ans.** $y = 60 - \frac{1}{15}x$

**1** $\frac{3}{5}$, $\frac{3}{5}x$

**Ans.** $y = 18 - \frac{3}{5}x$

**2** $y = 12 - \frac{6}{5}x$

**How to Solve**  The length of the incense stick is proportional to the time.
Because the stick becomes 6 cm shorter in 5 minutes,
it becomes $\frac{6}{5}$ cm shorter in 1 minute.
Therefore, it becomes shorter $\frac{6}{5}x$ cm after $x$ minutes.

**3** $y = 15 - 6x$

**㉟ Linear Functions 5 (Pattern 2)** pp 166,167

**★ TRY!**

(1) $y$, $b$

(2) 25, 31, 2, 15

(3) 2, 15, 11

**Ans.** 37 cm

**1** linear function, 9000, 12500,

$$\begin{cases} 200a + b = 9000 \\ 300a + b = 12500 \end{cases}$$

If solving these two simultaneous equations,
$a = 35$, $b = 2000$
Therefore, $y = 35x + 2000$

**Ans.** $19,500

② Assume the cost of making $x$ yearbooks is $\$y$.

Since $y$ is linear function of $x$, it can be expressed as $y=ax+b$.

$$\begin{cases} 60a+b=1900 \\ 75a+b=2300 \end{cases}$$

If solving these two simultaneous equations,

$a=\dfrac{80}{3},\ b=300$

Therefore, $y=\dfrac{80}{3}x+300$

Substituting $x=90$ in this equation,

$y=\dfrac{80}{3}\times 90+300$

$\quad =2700$ 　　　　　**Ans.** $\$2,700$

## 36 Linear Functions 6 (Pattern 3) 　pp 168, 169

★ **TRY!**

(1) 20

(2) 20, 0, 3, $-\dfrac{3}{40}$, 3

(3) $\dfrac{9}{10}$ 　　　　　**Ans.** $\dfrac{9}{10}$ km

① 5, $y=-\dfrac{1}{14}x+10$ 　　**Ans.** $\dfrac{15}{2}$ km

[Also, 7.5 km]

② The point 150 minutes after Roy's departure from the foot is after he departs the mountain cottage.

Since the graph is a straight line passing through the points $(80, 5)$ and $(180, 0)$, and the slope is $-\dfrac{5}{100}=-\dfrac{1}{20}$.

The equation representing the relationship between $x$ and $y$ is $y=-\dfrac{1}{20}x+b$.

Substituting $x=180$, $y=0$ into the equation, $b=9$.

Therefore, $y=-\dfrac{1}{20}x+9$

Substituting $x=150$ into the equation, $y=\dfrac{3}{2}$

**Ans.** $\dfrac{3}{2}$ km

[Also, 1.5 km]

## 37 Linear Functions 7 (Pattern 4) 　pp 170, 171

★ **TRY!**

(1) $\dfrac{1}{15}x$

(2) 6

(3) time

(4) 45, 3

(5) 45, 3 　　**Ans.** Alan's equation $y=\dfrac{1}{15}x$

**Ans.** Madison's equation $y=\dfrac{1}{5}x-6$

**Ans.** The time 10:45

**Ans.** The distance 3 km

① 4, 4, 50, $\dfrac{2}{25}$ 　**Ans.** Beverly's equation $y=\dfrac{2}{25}x$

4, 4, 40, $\dfrac{2}{15}$, $\dfrac{2}{15}$, $-\dfrac{4}{3}$

**Ans.** Wayne's equation $y=\dfrac{2}{15}x-\dfrac{4}{3}$

② Considering Beverly's equation $y=\dfrac{2}{25}x$ and Wayne's equation $y=\dfrac{2}{15}x-\dfrac{4}{3}$ as simultaneous equations,

$\dfrac{2}{25}x=\dfrac{2}{15}x-\dfrac{4}{3}$

$x=25$

Substituting the value into $y=\dfrac{2}{25}x$,

$y=2$ 　　　　　**Ans.** Time 9:25

**Ans.** Distance from their house 2 km

## 38 Linear Functions 8 (Pattern 5) 　pp 172, 173

★ **TRY!**

(1) 6, $x$, $4x$

(2) 14, 6, 24

(3) 20, $x$, 80, $4x$

**Ans.** On to and along side AB $y=4x$

**Ans.** On to and along side BC $y=24$

**Ans.** On to and along side CD $y=80-4x$

① 4, $x$, $\dfrac{7}{2}x$, 11, 4, 14, 15

$$\boxed{\begin{aligned} y&=\dfrac{1}{2}\times 7\times(4+7+4-x) \\ &=\dfrac{105}{2}-\dfrac{7}{2}x \end{aligned}}$$

**Ans.** On to and along side BA $y=\dfrac{7}{2}x$

**Ans.** On to and along side AD $y=14$

**Ans.** On to and along side DC $y=\dfrac{105}{2}-\dfrac{7}{2}x$

② Rectangle ABPD is trapezoid.

The formula for the area of a trapezoid is $\dfrac{1}{2}\times(a+b)\times h$ where $h$ is the height, and $a$ and $b$ are the lengths of the parallel sides.

When point P moves on side BC,

$0\le x\le 8$ and hence $y=\dfrac{1}{2}\times(x+8)\times 6=3x+24$

When point P moves on side CD,

$8\le x\le 14$ and hence $y=\dfrac{1}{2}\times\{(8+6-x)+6\}\times 8=80-4x$

**Ans.** On to and along side BC $y=3x+24$

**Ans.** On to and along side CD $y=80-4x$

## 39 Linear Functions Review 1 　pp 174, 175

① (1) **Ans.** Plan A $y=0.30x+16$

**Ans.** Plan B $y=0.40x+9.80$

(2) $\begin{cases} y=0.30x+16 \\ y=0.40x+9.80 \end{cases}$

$x=62$ 　　　　　**Ans.** 62 minutes

(3) Plan A $0.30\times 50+16=31$

Plan B $0.40\times 50+9.80=29.80$

**Ans.** Plan B, Plan B is $\$1.20$ cheaper than Plan A

**2** (1) $80 \times 5 = 400$
$y = 1200 - 400 = 800$  **Ans.** 800 m
(2) $1200 \div 80 = 15$  **Ans.** $0 \leq x \leq 15$
(3) When $x = 0$, $y = 1200$
From (1), when $x = 5$, $y = 800$
Substituting the values of $x$ and $y$ into $y = mx + b$
$\begin{cases} 1200 = 0 + b \\ 800 = 5m + b \end{cases}$
$b = 1200$, $m = -80$  **Ans.** $y = -80x + 1200$

**3** (1) $AP = 2 \times 2 = 4$ (cm)
$y = \frac{1}{2} \times 4 \times 8 = 16$  **Ans.** $y = 16$
(2) $y = \frac{1}{2} \times 2x \times 8 = 8x$  **Ans.** $y = 8x$
(3) $y = \frac{1}{2} \times 8 \times 6 = 24$  **Ans.** $y = 24$

**4** (1) The figure shows that Laura is 8 km from City A in 2 hours.
  **Ans.** $y = 4x$
(2) The figure shows that Laura took a break between 2 and 3 hours after leaving City A.  **Ans.** 1 hour
(3) Find the value of the slope of a straight line passing through two points (3, 8) and (5, 14).
Substituting the values of $x$ and $y$ into $y = mx + b$
$\begin{cases} 8 = 3m + b \\ 14 = 5m + b \end{cases}$
$m = 3$, $b = -1$
Therefore, $y = 3x - 1$  **Ans.** 3 km/h

## (40) Linear Functions Review 2    pp 176, 177

**1** (1) **Ans.** $y = -\frac{1}{3}x + 6$
(2) Substituting $y = 0$ into $y = -\frac{1}{3}x + 6$, $x = 18$
  **Ans.** (18, 0)
(3) Let $k$ be the $x$ coordinate of point P.
Since the areas of $\triangle AOP$ and $\triangle BPC$ are equal,
$\frac{1}{2} \times k \times 5 = \frac{1}{2} \times (18 - k) \times 2$
$k = \frac{36}{7}$  **Ans.** $\frac{36}{7}$

**2** (1) Since $y$-intercept is 8,
Substituting $x = 4$ and $y = 0$ into $y = mx + 8$
$0 = 4m + 8$
$m = -2$  **Ans.** $y = -2x + 8$
(2) Substituting $x = k$ into $y = -2x + 8$
  **Ans.** $-2k + 8$
(3) The equation of the straight line AB is $y = x + 8$.
Substituting the $y$ coordinate value $(-2k + 8)$ of point R into $y = x + 8$,
$-2k + 8 = x + 8$
$x = -2k$
Therefore, $SP = k - (-2k) = 3k$
From $QP = SP$, $-2k + 8 = 3k$
$k = \frac{8}{5}$
$SP = 3 \times \frac{8}{5} = \frac{24}{5}$  **Ans.** $\frac{24}{5}$ cm

**3** (1) **Ans.** $y = 2x + 2$
(2) **Ans.** $y = -x + 8$
(3) Find the $x$ coordinate of the intersection point.

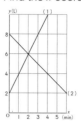

(4) **Ans.** 2 minutes later

**4** (1) As Timothy walked over 1200 minutes for 20 minutes,
$1200 \div 20 = 60$  **Ans.** 60 m/min
(2) As Jason walked over 1200 minutes for 30 minutes,
$1200 \div 30 = 40$  **Ans.** 40 m/min
(3) Find the value of $x$ from the equation of two straight lines.
$\begin{cases} y = 60x \\ y = -40x + 1200 \end{cases}$
$x = 12$  **Ans.** 12 minutes later

## (41) Review 1    pp 178, 179

**1** $0.50  45 stamps
$0.80  35 stamps
**2** Large 4 bags
Small 8 bags
**3** Jeffrey $11.40
Ryan $10.20
**4** A 8%
B 5%
**5** Boys 300
Girls 250
**6** 56 marigolds

## (42) Review 2    pp 180, 181

**1** Run 9 km
Walk 1 km
**2** 400 m
**3** Pencil $0.42
Eraser $1.05
**4** 37
**5** Morning 12 fish
Afternoon 11 fish
**6** Walk 1,200 m
Run 1,600 m

## (43) Review 3    pp 182, 183

**1** $m$ 19
$n$ 60
**2** Adult $13
Child $7
**3** From home to bookstore
480 m
From bookstore to station
720 m
**4** Metal $1.50
Paper $0.80
**5** 4 oranges 12 bags
6 oranges 27 bags
**6** Boys 246
Girls 286